Renewable Energies with Energy Storage

Renewable Energies with Energy Storage

Winston (Win) Stothert

Copyright © 2011 by Winston (Win) Stothert.

Library of Congress Control Number:		2011904590
ISBN:	Hardcover	978-1-4568-9108-4
	Softcover	978-1-4568-9107-7
	Ebook	978-1-4568-9109-1

All rights reserved. No part of this book may be reproduced or transmitted in any form or by any means, electronic or mechanical, including photocopying, recording, or by any information storage and retrieval system, without permission in writing from the copyright owner.

This book was printed in the United States of America.

To order additional copies of this book, contact:
Xlibris Corporation
1-888-795-4274
www.Xlibris.com
Orders@Xlibris.com
93160

I am grateful to my wife, Sylvia "Hollie" Stothert for her unwavering support and constant encouragement. I appreciate the assistance with graphics by my darling granddaughter, Jessie McGrath and by my friend Jane Alley. Our families have shared my interests with moral support throughout the development of this works.

Contents

Statement Regarding Conclusions ix
Observations xiii

Chapter 1:   Energy, Yesterday, Today and Tomorrow 1
Chapter 2:   Current Hydrogen Production
              Method and Costs 14
Chapter 3:   Renewable Wind Energy Sources and Costs 20
Chapter 4:   Renewable Solar Energy Sources and Costs 49
Chapter 5:   Hydrogen by Electrolysis from
              Renewables, with Costs 68
Chapter 6:   World's First Private Commercial Hydrogen
              Production Plant by Electrolysis 79
Chapter 7:   Hydrogen from Other Sources 86
Chapter 8:   Carbon Dioxide Emissions and Credits 89
Chapter 9:   Distributed Power by Hydrogen Fuel Cells 102
Chapter 10: Production Potential for Hydrogen from
              Renewables 116
Chapter 11: Wind Energy to Hydrogen to
              Ammonia Fertilizer 137
Chapter 12: Motor Vehicles and Emissions 142
Chapter 13: Gasoline Consumption and
              Hydrogen Substitution 150
Chapter 14: Alternative Storage of Variable
              Renewable Energies 156
Chapter 15: Water Used – Water Saved 178
Chapter 16: Observations and Conclusions 190

Appendix 203

Conclusions based on opinions can be a detriment to technological advancement.

Conclusions based on facts gained from technological advancement can lead to practical utilization for the betterment of humanity.

Our world has three interconnected crises:
climate, energy and water

Some of the answers to these crises are here.

When, inevitably, world energy demand exceeds oil and gas supplies, then these answers will be very acceptable.

Unless . . .

We may have reached the point where greenhouse gasses from fossil fuels have brought climate to the disastrous state of no-recovery—beyond the tipping point.

On a financially acceptable basis, it is possible to eliminate the need for overseas oil imports and reduce greenhouse gas emissions to an acceptable level.

It is also possible to eliminate the need for any crude oil or coal used for energy.

A combination of renewable energies used directly in the grid, with some converted by electrolysis to hydrogen, and natural gas turbine—generators using compressed air energy storage systems—

will feed electric power from these storage systems to the grid and fuel vehicles, achieving these objectives.

Humans began storing energy when fire was invented; since then, hundreds of trillions of dollars have been invested in energy storage.

Renewable Energies For The World

OBSERVATIONS

It has been technologically proven that renewable energy sources can be developed in a socially responsible manner to supply all of the world's energy requirements, without production of greenhouse gases. It can eliminate the need for fossil fuels.

Cost of renewable energy is often compared to the cost of electricity from existing fossil-fueled nuclear power and stored hydro-generating plants built decades past. Cost of power from new carbon-neutral fossil-fueled power plants will be equal to or at a higher cost than from renewables of wind, solar, and biomass. Wind and solar have no future cost increases due to inflation, to which the other sources are subject. Cost of any new generation systems, which are required to meet growth in demand, will cost more due to decades of inflation. These costs melded with the cost of the older plants will inevitably increase the average cost to the consumers whether from historic methods or from the new world energy.

Today's major established sources of renewable energy are wind, solar-thermal, solar-photo-voltaic, hydro, run-of-river hydro, biomass, and geothermal. Other sources—including tidal, wave, and solar-hydrogen—are developing.

Energy from renewables is variable by nature, except for biomass and geothermal; therefore it requires a means of storage of the energy to be drawn on during periods when the renewable energy is not available. They can become a firm supply by use of hydrogen for storage, much like utilizing large hydro systems. An alternative to large-scale use of hydrogen for storage is to use the renewables to operate compressors and store compressed air in caverns with recovery of the energy through gas turbine generators, which then produce only one-third of the greenhouse gases compared to conventional gas turbine systems and a small fraction of the emissions from coal-fired plants.

Technology for recovery of renewable energy from its various sources has advanced exponentially in the past two decades and is accelerating.

Achieving the replacement of fossil fuels by renewable energy needs informing the public with clear statements of fact, followed by effective public policies and carefully directed incentives. Comprehensive estimates of adaptation and benefits are available but need better dissemination.

Hydrogen fueling stations for vehicular use can have the hydrogen produced on site with today's proven technology, from natural gas or by electrolysis. No transport of the hydrogen is necessary, and either energy source already has an existing distribution network.

The energy efficiency of fuel-cell vehicles with hydrogen is twice that of those using gasoline. Projections are made that this efficiency advantage will be increased to two and one-half times. It had been considered that one kilogram of hydrogen was required to replace one gallon of gasoline. With the better efficiency of fuel-cell vehicles, one kilogram of hydrogen is equivalent to two gallons of gasoline.

Distributed Power Centers (DPCs) can be installed with today's proven technology at electrical power load centers with hydrogen produced either from natural gas or electricity, with the hydrogen stored and converted through fuel-cells to electricity to meet high-value peak demands. Sufficient DPCs can avoid the investment in new power generation, new transmission lines, new distribution facilities, and conventional emergency standby power plants by eliminating the need of these additional facilities to supply when only peak power is needed or in case of blackouts.

Compressed air energy systems are an attractive alternative for major quantities of electric power storage.

Carbon, as in carbon dioxide, is an element of the earth, which is neither created nor destroyed. The carbon contained in fossil fuels came originally from the earth's atmosphere over hundreds of millions of years. It might be argued that returning it to the atmosphere by combustion of the fossil fuels will have no adverse effects. However, is it necessary to consider that carbon removed from the atmosphere over hundreds of millions of years but returned to the atmosphere in a short period of one hundred years allows the Earth to accept it without adverse effects?

This text guides to the conclusions that

- renewable electricity can be produced in sufficient quantities, at a reasonable combination of investment and operational costs, together with electricity production from natural gas combined cycle turbine systems to supply all of the world's needs for many generations. The continued use of the fossil fuel, natural gas with its comparatively much-reduced GHG emissions, is a practical compromise in the adjustment to a GHG-free economy.
- Electricity and hydrogen from renewables can fuel all of the world's transport vehicles without the need for gasoline from crude oil.
  Natural gas, a fossil fuel, is a limited quantity and nonrenewable. The ultimate objective must be to provide all the world's energy needs from renewable sources. That would meet an aim of the Bruntland Commission of the United Nations: "Sustainable development is development which meets the needs of this generation without sacrificing the ability of future generations to meet their needs."

Chapter 1

ENERGY

Yesterday, Today, and Tomorrow

WAVES OF INNOVATION

No. 1. The Steam Engine

In the mid-1800s, the steam engine was used as a stationary power source to begin replacing wind and hydro power, which did not always provide the power when needed. It was uniquely different than those sources in that it could provide power whenever it was needed and while moving. This resulted in steam locomotive powered trains utilizing a fossil fuel, coal, as the energy source. For the first time, man was able to travel faster than 30 miles per hour, the speed of a horse. It was limited in its routes. Production rates were increased by replacing man or animal with the steam engine, resulting in a higher standard of living. Fossil fuel was the energy source.

No. 2. Electricity

At the start of the 1900s, electricity was harnessed as a new source of energy from hydro and from steam engines driving turbine generators. Power became more readily available without having to be adjacent to a steam, hydro, or wind source. Production was enhanced and the standard of living increased. Fossil fuel was one of the energy sources; renewable energy in the form of hydro was the other. Electricity was harnessed with batteries to buggies for the first electric vehicles.

No. 3. The Internal Combustion Engine

Also at the start of the nineteen hundreds the internal combustion engine was invented and quickly found its niche in powering passenger and freight vehicles. Man was now able to travel up to 100 miles per hour and faster. At the same time this flexible form of power for movement of people and goods improved the standard of living. Fossil fuel was the energy source.

No. 4. The Airplane

Following the first quarter of the nineteen hundreds, development of the airplane increased the speed at which man could travel to 300 miles per hour and then to exceed the speed of sound. The airplane provided travel for man on a global basis for business and pleasure. It contributed further to an increased standard of living. Fossil fuel was the energy source.

No. 5. Satellites and Space Travel

Space travel later in the nineteen hundreds again increased the speed at which man can travel, astronomically; and the resulting technological discoveries and developments needed to achieve this were applied to the conventional industries, again leading to a higher standard of living. *Hydrogen*, not fossil fuels, was the energy source which made space travel possible, combined with oxygen.

No. 6 Renewable Energy

Economic growth in the developing countries and continuing increase in demand for energy to meet ever-higher standard of living in developed countries has caused an explosive growth in demand for energy. This has created the dual crises in climate and energy, which now require an incredibly major and rapid response to meet the energy needs which will not be available from the finite supply of fossil fuels, other than coal, and to reverse the accumulation of greenhouse gases, which many claim will otherwise destroy our life on this planet.

Standard of Living and Energy:

Every time that the speed at which man could travel and the standard of living was increased, the consumption of energy increased. The main sources of this energy have been the fossil fuels: oil, natural gas, and coal. Some power has come from hydro and nuclear sources.

Finite Supply of Fossil Fuels:

At the same time that the standard of living has increased, the world's population has multiplied astronomically. The demand for fossil fuels has risen. However, the supply of fossil fuels is finite. As the standard of living rises in the heavily populated and previously underdeveloped parts of the world, meeting their needs of fossil fuels will require a sharing of those fuels now almost fully utilized by the smaller developed world.

Adjustments will be necessary *without a choice*, except for all nations, to maximize their efficient use of energy and use the several methods of developing new sources of energy—*only renewable energies offering the total solution.*

Energy Consumption in the Developing World:

The standard of living of people in the underdeveloped regions of the world can be measured in several ways, one of which is the average speed of travel available to them compared to those in the developed world, generally less than the speed of a horse. Another is their availability and consumption of fossil fuels, almost nil.

As some sectors of the underdeveloped world are advancing and increasing their standard of living, they require an increasing amount of fossil fuels, the most versatile being crude oil. The combination of this increase in world demand for crude oil, together with the fact that most of the easy oil has been discovered and is now at a depleting production rate, has resulted in major increases in the price of the oil. The combination of rapidly increasing world demand and the increasing cost of new crude oil sources can be expected to continue, possibly at an exponential rate. At the same time, the demand from the developed world has not leveled off. A world energy crisis has arrived.

All forms of energy used by humans is now 15 terrawatts (TW) (e) annually (source: Wikipedia 2008). One terrawatt (TW) is 10^{12} watts and (e) refers to the equivalent in electricity measurement. There are predictions that this will increase, by 2050, to 30 terrawatts annually. Other sources refer to the current total world *electric power consumption* at about 13 terrawatts. (Elsevier, December, 2008, Muradove & Veziroglus.)

Fossil fuel consumption in the United States is currently in the order of 3.5 TW (e).

In terms of nuclear power (GHG free in operations), the equivalent power required to replace today's U.S. fossil fuel consumption would involve the equivalent of 3,500 plants of 1 gigawatt (GW) (10^9 watts) capacity each (there are now 103 nuclear plants in the United States), and there is some doubt as to whether sufficient uranium could be found to supply them. In some regions there will not be enough cooling water.

Alternative Energies:

Development of alternative sources of energy has always been in progress but not with the effort that is now known to be essential and immediate. Hydro sources continue to be exploited, but the opportunities are limited, with most of the practical ones already in operation. Canada is the world's largest producer of hydroelectricity. China is developing new hydroelectric generation facilities at breakneck speed. British Columbia has a major underdeveloped resource for run-of-river nonstored hydro in the order of 11,000 MW (Ref. Wikipedia). Natural gas, a fossil fuel, is an alternative with limited potential as the supply is only matching current demand; and most existing sources are operating at a depleting rate. New major sources in North America are being developed; however, again, the easy fields have been developed; and the new ones come at a much higher cost of production. Costs, safety factors, and other considerations limit the transportation of natural gas across oceans from available supply to areas of demand.

Coal:

Coal, a fossil fuel, is more abundant than oil or natural gas. However, again, the easiest low-cost coal has been mined and new sources cost considerably more. Coal is a bulk commodity, and moving it from source to point of use is costly. In today's world of concern with greenhouse gases, it is the greatest emitter of the three main fossil fuels, a condition which can only be rectified if the coal is gasified at a power plant and a concentrated form of carbon dioxide gas can be sequestered in subterranean safe storage. There has been concern expressed as to the risk of subterranean storage with a sudden leak. There was an incident of sudden carbon dioxide release from an old volcano lake in Africa, which asphyxiated 3,000 people. There is also some doubt expressed as to the availability of sufficient subterranean space for storage of the quantities of carbon dioxide which would have to be stored on a cumulative basis if much of the future demand for electricity were to be met by coal-fired plants.

Renewable Energies:

There is great promise from renewable energy sources, primarily wind and solar. Development of these has been at an astounding rate recently, reducing costs to a more competitive level. Their current disadvantage is their variability due to being dependent on nature. A means of making them more acceptable is to convert them from their variable nature to provide firm energy. This can be accomplished by using the electricity for electrolysis of water, producing hydrogen gas which is a form of energy that can be stored and used on demand directly as a vehicle fuel or by reconversion to electric power. At least ten other means of storing electric energy have been established or are in the development stage. Some of these have the advantage of quick response but limited capacity. For large quantities and seasonal storage, the most promising are compressed air energy storage systems (caes) and hydrogen.

Development of renewable energy has become a worldwide objective to meet the growing demand for electrical power and vehicle fuels, which are exceeding the availability of fossil fuels (other than coal), and which is essential for reduction of greenhouse gases. Sources of renewable energy have been proven to be sufficient to supply all of the world's energy needs. However, as noted, wind and solar—along with other renewables, such as wave and tidal sources of renewable energies—are variable and require some provision to make them available on a demand basis. Hydrogen and CAES are two of the major methods of providing renewable energy on demand.

Bioenergy is a proven and developed firm supply source of renewable energy and is considered to be carbon dioxide neutral. There are a number of sources, but they are limited in availability except for the potential of algae. Algae development as a source of bioenergy has great potential; and a significant advantage, especially when compared to crop-grown biofuels, is that it requires only a small fraction of the land space.

Dual Crises of Climate Change and Reliance on Fossil Fuels with Replacement by Renewables:

There are three main avenues being considered for a carbon-neutral supply of energy. These are from carbon-removed fossil fuels, nuclear energy, and nature's renewable energy sources. Only the latter two will reduce or eliminate the dependence on fossil fuels.

Elsevier Publication, December, 2008 study "Green Path from Fossil-based to Hydrogen Economy" suggests 10 TW of carbon-free power will have to be produced by 2050 in order to "keep our planet healthy."

The hydrogen economy, by electrolysis, has been shown to be sufficient from renewable sources to supply all the world's vehicle fuel and to replace the use of fossil fuels for supply of electric power.

Global primary power consumption is currently about 13 TW.

Current Hydrogen Production:

Current production of hydrogen is over 90% from fossil fuels, primarily by steam reforming methane (natural gas). The Elsevier 2008 report previously noted indicates 9.5 kg of carbon dioxide (CO_2) is produced with each 1 kg of H_2 (0.44 Nm^3 CO_2 / Nm^3 H_2).

It also projects the cost of carbon capture and sequestration (CCS) could be in the order of $20/tonne of CO_2. The methane-reforming process for the production of hydrogen results in excessively high emissions of carbon dioxide to the atmosphere.

Note: For thermal coal-fired power plants, other studies have indicated the full cost, including the cost of the thermal process which will provide a concentrated carbon dioxide exhaust and sequestration, could be in the order of $100 per tonne of CO_2.

Hydrogen—The Element

Hydrogen is the major element in the developed world. Its current production is a major contributor to greenhouse gases, utilizing fossil fuel, and is therefore nonsustainable.

Clean hydrogen can be produced by electrolysis of water using electricity from renewable energy sources. This is sustainable.

- It is the lightest of all known elements, with an atomic weight of 1.0008.
- It is a flammable, colorless, and odorless gas.
- It is the most common element at 80% of the universe.
- It is part of every living thing.

Examples of some of the most common chemical compounds, all containing hydrogen, are:

H_2O	Water
H_2SO_4	Sulphuric acid, the most common acid and a commonly used catalyst.
CH_4	Methane, natural gas.
C_6H_6	Benzene, a most common organic compound.
C_2H_5OH	Ethyl alcohol
NH_3	Ammonia, a fertilizer currently using the second largest quantity of hydrogen
$C_{12}H_{22}O_{11}$	Sugar

Gasoline is 15% to 16% hydrogen.
The list is endless.

The higher heating value (HHV) of hydrogen at 25 °C is 142 MJ/kg (1.055 MJ = 1,000 btu), and 142 MJ/kg = 135,000 btu/kg. For gasoline, the average is 130,000 btu per gallon.

> MJ—megajoules; Joules are a measure of energy. Mega is million.
>
> btu—British thermal units; a measure of energy

Hydrogen's Safe Use

Hydrogen codes and standards are being developed to provide the information needed to safely build, maintain, transport, and operate hydrogen and fuel cell systems and facilities. Codes ensure uniformity of safety requirements and provide local officials and safety inspectors with the information needed to certify hydrogen systems and installations. Hydrogen can be used as safely as today's common fuels when guidelines are observed and users understand its behavior. (National Renewable Energy Laboratories—DOE—U.S.) (NREL)

Hydrogen Usage—World Wide

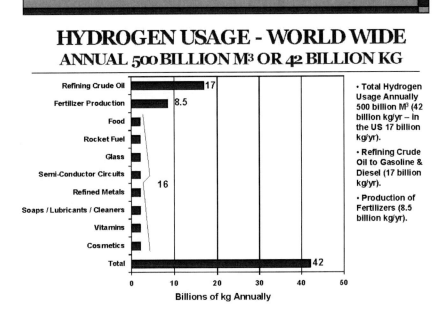

Hydrogen by Electrolysis

Jules Verne in 1874 wrote in *The Mysterious Island*, "I believe that water will one day be employed as a fuel, that hydrogen and oxygen will constitute it, used singly or together, will furnish an inexhaustible source of heat and light."

Electrolysis—The Theory

Michael Faraday (1791-1867) discovered the quantitative relations between the amount of electricity that passes through a solution and the quantity of matter separated at the electrodes.

His first law is "The quantities of substances set free at the electrodes are directly proportional to the quantity of electricity which passes through the solution."

His second law expresses the fundamental relation between quantities of different substances liberated at the electrodes by the same quantity of electricity. It is "The same quantity of electricity sets free the same number of equivalents (equivalent weights) of substances at the electrodes." This is a directly proportional relationship.

A quantity of electricity, 96,459 colombs (a colomb is an ampere-second) is called a faraday. If one faraday of electricity is passed through an electrolytic conductor, 1 gram equivalent of some substance will be liberated at each electrode.

In practice, more than 1 faraday is needed for the liberation of a gram equivalent of a substance. This is due to side reactions, mechanical losses of the products of the reaction, secondary reactions at the electrodes, current leaks, and losses in the form of heat. The ratio of the theoretical to the actual quantity of current used is the current efficiency.

Source: John Perry, Chemical Engineers Handbook, McGraw-Hill

Electrolysis Efficiency:

Electricity to hydrogen conversion efficiency by electrolysis is reported to be 63%. Compression of the hydrogen for economic storage has an efficiency in the order of 90%. As noted in chapter 3, the NREL report (Boulder, Colorado) uses a conversion from electricity to hydrogen of 58.8 kilowatts/kilogram (kWh/kg). One kg of hydrogen has a higher heat value (HHV) of 142 MJ (megajoules) or 135,000 btu (British Thermal Units). One kilowatt hour (kWh) of electricity = 3, 413 btu/hr. The energy input of 58.8 kWh/kg is 58.8 kWh/kg x 3,413 btu/kWh = 200,684 btu/hr. The efficiency used by NREL for electrolysis is then 135,000/200,684 = 67.3%. (NREL report states they used 66.3%). NREL report the average efficiencies of some of today's electrolyser manufacturers—such as Proton, Avalence, Teledyne, Stuart, and Norsk-Hydro—is 58.8 kWh/kg. The potential quantities of hydrogen from renewables in the United States reported by NREL used these efficiencies and do not include the loss factor for compression.

A further measure of the efficiency of an electrolyser can be based on the theoretical reaction in which the thermodynamic amount of energy used in production of 1 kg of hydrogen is 39 kWh. This consumption over the reported efficiencies, 39/58.8, indicates an efficiency of 66%. There are energy requirements for ancillary equipment together with heat loss.

Electrolysers and MIT:

MIT's Henry Dreyfus Professor of Energy Dr. Daniel Nocera announced in 2008 a new patented method of converting the sun's energy to hydrogen and then through fuel cells to electricity. It is indicated a chemical catalyst (cobalt and potassium phosphate) is added to water, and a current of electricity converts the water to hydrogen and oxygen. The claim is made that the cost of producing the hydrogen will be only a few percent of the current costs using today's electrolysers.

Wind Increase in Renewables by 2050

The International Energy Agency (IEA) has issued a report entitled "Deploying Renewables: Principles for Effective Policies." It estimates that 50% of the world's energy must come from renewables by 2050 in order to reduce greenhouse gases by 50%. The report shows potential growth of the various renewables from 2000 through 2050 from just over 3,000 terrawatt hours per year (TWh/yr) to 19,000 TWh/yr; an addition of 16,000; an average of 320 TWh/yr. One terawatt is 10^{12} watts.

Wind energy over the 50 years is projected to supply one-half of the renewables or 8,000 TWh for an average addition of 160 TWh per year. This average *annual* addition of wind energy would require:

> 160 x 10^{12} watt-hours/yr / 8,760 hrs/yr / 10^6 watt-hours per megawatt hour = 18,260 MW, annually, at full capacity.

With a capacity factor (CF) for the wind energy of 0.30, the installed wind farm capacity required would be 18,260 MW/0.30 CF = 61,000 MW per year of wind-generation facilities.

At an average capital cost of $2 million per MW, the *annual* investment worldwide would be $122 billion. Add 25% to this for transmission, and the total annually would be $152 billion.

The world installation of wind energy in 2009 was 37,500 MW, 61% of the IEA advisory annual average for 50 years.

Crude Oil Replacement by Hydrogen

One "barrel" of crude oil is 42 gallons (Imperial). Refining by distillation and catalytic cracking will convert 50% (depending on the quality of the crude) to gasoline, i.e., 21 gallons per barrel.

If all of the vehicles in the United States were converted to hydrogen, 69-billion kilograms of hydrogen would be required to replace the gasoline portion of 6.6 billion barrels of crude oil annually. See chapter 11.

A gallon of gasoline has, depending on quality, 130,000 btu/gallon. Hydrogen has an energy equivalent of 135,000 btu/kg. A kilogram of hydrogen is sometimes referred to as equivalent to the energy of one gallon of gasoline (gge). However, due to the relative efficiency of conversion in the vehicle, one kilogram of hydrogen will provide a travel distance twice that of one gallon of gasoline.

Referring to chapter 3, if one-half of the wind energy potential (600,000 MWh/hr) of the United States was converted by electrolysis to hydrogen to be used in vehicles, the hydrogen production would be 89 billion kg/year (with target electrolyser efficiency of 63%). At the target 63%, electrolyser efficiency, the electricity usage would be 58.8 kWh/kg (see chapter 11). The 600,000 MWh/hr / 58.8 kWh/kg would convert to 89 billion kilograms of hydrogen annually, which would have required the equivalent of 8.5 billion barrels of crude oil (for 21 gallons of gasoline per barrel and 1 kg of hydrogen equivalent to 2 gallons of gasoline).

Note: NREL Report of February 2007 shows, in a socially responsible manner, that renewable energy in the United States has the potential to produce 1 billion tonnes annually of hydrogen (5.3 times the hydrogen requirement for the world's vehicles).

Distributed Power Centers (DPCs):

Distributed power centers generate electricity or store energy and convert it to electricity at the electric power demand centers instead of the present practice with most power generation remote from where it is most needed.

An investment in DPCs can reduce the peak generating requirements and defer new investment in generation, transmission, and distribution facilities.

Chapter 2

CURRENT HYDROGEN PRODUCTION METHOD & COSTS

Hydrogen Production Today

Production of most of today's hydrogen is by steam reforming of natural gas. Natural gas, combined with steam, in the presence of a catalyst, converts to hydrogen and carbon dioxide.

$$CH_4 + 2H_2O = CO_2 + 4H_2$$

The conversion, on a molecular basis, is:

$$(12 + 4) + (4 + 2 \times 16) = (12 + 2 \times 16) + 8$$

16 kg of methane yields 8 kg of hydrogen, omitting the efficiency factor.

44 kg of CO_2 are produced with 8 kg of hydrogen (a ratio of 5.5 to 1), again without applying the efficiency factor.

Applying a process efficiency of 85% indicates the yield of one kilogram of hydrogen would be obtained from 2.35 kilograms of natural gas.

Also adjusting for an 85% process efficiency, the quantity of carbon dioxide which would be emitted to the atmosphere would be higher.

6.5 kg of CO_2 is emitted to the atmosphere for each 1 kg of hydrogen produced. (5.5 kg. of CO_2 per kg of H_2 x 0.85% efficiency = 6.5 kg of CO_2 per kg of H_2). This is before any purification and before compression needed for storage and transportation.

Nonsustainability with fossil fuel. This is one of the world's largest emitters of carbon dioxide (greenhouse gas) to the atmosphere.

The process efficiency of hydrogen from steam-reformed methane (natural gas):

The process efficiency is in the order of 85% (ranges from 75% to 90%).

This hydrogen requires purification in order to use it in foods, fuel cells and some other applications. This is done by pressure swing adsorption (PSA), which also reduces the overall efficiency.

A third step is compression, generally to 5,000 or 10,000 pounds per square inch (psi), for economic storage and transportation. This again reduces the overall efficiency by about 10%.

The global efficiency of the process for purified, compressed hydrogen will be in the order of 70%, with 1 kg of hydrogen requiring 2.9 kg of methane. Life-cycle emissions for production of the methane and the reformer plant are not included. Note: Some reformer plants claim a global efficiency closer to 85%.

Potential for Sequestration of Carbon Dioxide from the Reforming Process, with Cap and Trade Credits:

The current method of providing over 90% of today's hydrogen is by steam-reforming natural gas (methane). As indicated by the preceding formulae and allowing for efficiency loss, there are 6.5 kilograms of carbon dioxide produced for each kilogram of hydrogen. Current world production of hydrogen is 42 billion kilograms. With 90% or more produced by steam-reforming methane, this would be 38 billion kilograms, and the emitted carbon dioxide would be 247 billion kilograms annually (247 million tonnes).

Most hydrogen produced by this method is from large central plants; some, especially those supplying fertilizer plants, are located near oil and gas fields. The carbon dioxide stream is concentrated with few other components, and there is an excellent opportunity of sequestering this greenhouse gas by compressing it and pumping it into subterranean caverns or to improve production from depleted oil wells. An opportunity for reduced cost is indicated from cap and trade versus the charge, which is expected from the carbon dioxide production with this method.

A pipeline is currently under construction in Central Alberta by Enchance Energy Inc. to carry a high concentration carbon dioxide stream from the large international fertilizer producer Agrium, at its Redwater plant, to depleted oil fields in the Clive area for injection. The intent is to improve recovery of oil from these fields and to expand the initial pipeline into a network carrying carbon dioxide for sequestration.

Cost Per Kilogram of Methane-Reformed Hydrogen:

This estimate is conceptually developed and *requires verification*. Cost of transportation is not included. Production costs are not readily available in the market place.

Cost of methane / mmbtu (1)	$4.00	$8.00	$14.00	$20.00	$32.00
Cost per kg of methane	$.19	$.39	$.68	$.97	$1.55
Methane 2.9 kg per kg H$_2$	0.55	1.13	1.97	2.81	4.50
Purification by PSA & 5,000 psi compression. Effy. 0.8% efficiency	0.11	0.23	0.39	0.56	0.90
Plant Charge	1.20	1.20	1.20	1.20	1.20
Cost of Hydrogen per kg	1.86	2.56	3.56	4.57	6.60
With carbon sequestration for new steam-reformed methane plants at up to $100.00 per tonne of CO$_2$. ($0.10 per kg and 6.9 kg of CO$_2$ per kg of H$_2$ at 70% effy.)	0.69	0.69	0.69	0.69	0.69
TOTAL COST OF HYDROGEN PER KG with Carbon sequestration	$2.55	$3.25	$4.25	$5.26	$7.29

Note (1). This table indicates the reliance of the cost of hydrogen produced by this method on the cost of natural gas.

World Production and Cost of Methane-Based Hydrogen

World production of hydrogen in 2005 was reported to be fifty million metric tonnes with an economic value of $135 billion (Wikipedia). Global hydrogen production currently is 48% from natural gas, 30% from oil, 18% from coal and 4% from water electrolysis.

135×10^9 / 50×10^9 kg = $2.70 per kg of hydrogen (economic value)

Hythane (Hydrogen added to Methane (Natural Gas)

In regions where hydrogen is produced by electrolysis from wind, solar, and other renewable energy sources and is served by natural gas pipelines, a significant addition of hydrogen to the methane can be permitted. The resulting mix can be used in lieu of 100% methane for its normal uses and, in case of vehicles, significantly reduces the greenhouse gas emissions. The existing network of natural gas pipelines can provide a practical and economical method of transmitting renewable electrical energy, converted to hydrogen, to the centers of energy demand.

CONVERSION DATA—HYDROGEN

Density of Hydrogen 1 Nm^3 = 90 gm. = 38 SCF (NTP)
1 kg of H_2 = 422 SCF = 11 Nm^3
Nm^3—cubic metre at normal temperature and pressure
SCF—standard cubic foot at normal temperature and pressure
NTP—normal temperature and pressure

CONVERSION DATA—NATURAL GAS

1 mm btu of natural gas = 1,000 SCF = 28.32 Nm^3 (standard cubic metres)
Natural gas density is 1.61 lb/Nm^3 = 0.73 kg.
1,000 SCF = 28.32 Nm^3 × 0.73 kg/Nm^3 = 20.67 kg/mmbtu or /1,000 SCF

Natural gas at $8.00 per mmbtu divided by 20.67 kg = $0.39 per kg.

Air Liquide S.A. New Hydrogen Plant

Industry News has reported among the large chemical industries that Air Liquide is planning a 120 million scf/day (3.6 million Nm3) hydrogen plant in Pasadena, Texas.

1 kg H_2 = 422 SCF.

120,000,000 SCF/day / 422 SCF/kg H_2 x 365 days/yr = 104,000,000 kg/yr H_2. or 104,000 tonnes.

Hydrogen from Electrolytic Chemical Plants

Electrolytic chemical production plants, primarily chlor-alkali facilities, produce hydrogen as a waste stream. It can be cleaned up by removal of moisture and traces of chlorine, compressed, and supplied either for fuel-cell vehicles or for generation of electricity through fuel cells or internal combustion engines. This is not an insignificant source of hydrogen, although finite, and the recovery costs are reasonable.

Chapter 3

RENEWABLE WIND ENERGY SOURCES AND COSTS

TAKING ALL ASPECTS INTO ACCOUNT,

THE COST

IS LESS THAN ANY OF THE ALTERNATIVES.

Wind Energy Sources and Costs

Introduction

An overview of the total primary energy use and total electricity energy use on a world scale, with projections, is necessary to relate to the planned projections of wind energy developments worldwide. As the United States is used as a basis for evaluating the potential of renewable energies due to its high level of utilization, its current and projected energy consumption and demand is also noted.

All Forms of World Energy

All forms of energy used by humans in 2008 was 474×10^{18} joules (exajoules). (Ref. Wikipedia.) Converting this to kilowatt hours equivalent: Note 1.

> 474×10^{18} joules per yr. / 3.6×10^6 joules per kWh = 132×10^{12} kWh/yr (e). or 132×10^{15} watt hours/yr (e),. (petawatt hours/yr).

The average consumption rate would be

> 132×10^{15} watthours/yr. / 8,760 hrs/yr. = 15×10^{12} watt/hr(e). (15 terrawatts/hr(e).

All forms of energy include those used for building heating and cooking, for transportation, and industrial processes.

There are predictions that this will increase by 2050 to 30 terrawatts (e).

The U.S. DOE Energy Information Administration Outlook 2009 reported the world total primary energy consumption (all forms of energy) in 2006 at 472 quadrilllion btu (472×10^{15}). The electrical *equivalent* of this would be

472×10^{15} btu/yr / 3,413 btu / kWh = 138×10^{12} kWh/yr (e). (138×10^{15} watt hours/yr(e) (138 petawatt hours/yr (e).

This correlates with Wikipedia, showing a slight reduction from 2006 to 2008.

The average demand in electrical *equivalent* would be

138×10^{12} kWh/yr / 8760 hrs/yr = 15.8×10^9 kWh/hr (e) (or 15.8 terawatts per hr).

Note 1: See appendix for conversions and electrical units.

World Electricity Generation

The DOE/EIA International Energy Outlook 2009 reported the total world electric net power consumption in 2007 at 17,109 billion kilowatt hours (17 petawatt hours). The *average* rate of consumption would be

17×10^{15} watt hours/yr / 8,760 hrs/yr = 1.9×10^{12} watts (1.9 terawatts or 1,900 gigawatts) (approximately two million megawatts)

Note: The peak demand would be in the order of twice the average or 4 terawatts (4,000 gigawatts).

The DOE/EIA reports for 2010 electrical generation in the United States to be 4,000 billion kWh (4 petawatt hours) *approximately* 4 / 17 = 24% of the world production.

A projection of world electric power load in 2050 of 37,000 TWh was reported in 2007 by Asko Vuorinen, author, and Ekoenergo

Oy, publisher, Finland. The author does provide two scenarios, with this being the lower quantity. Based on 17,000 TWh in 2007, the average increase annually would be

37,000 TWh—17,000 TWh / 43 years = 465 TWh per year.

A projection of the world net electricity generation provided by the U.S. DOE/EIA International Energy Outlook 2010 follows. The U.S. projection through 2035 is higher than that of Asko Vuorinen, who has also developed a higher projection scenario.

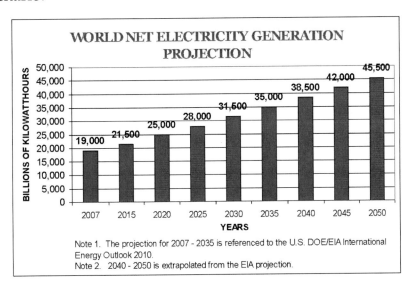

The following chart provides a prediction by the United States Department of Energy of the potential focus on renewable energies and natural gas as the two main sources of new generation during the 2008-2035 period in the United States.

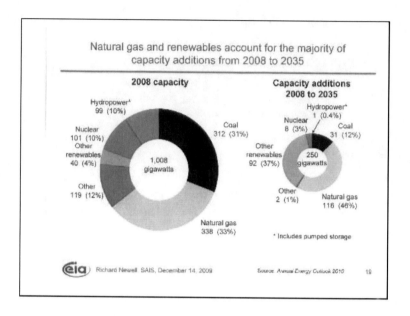

Notes to Chart:

1. Other renewables, excluding hydro, increase from 40 gigawatts and 4% of the total generation capacity in 2008 by 92 gigawatts and 37% of the added quantity.

2. Natural gas represents 46% of the added generating capacity with 116 gigawatts.

3. The two leading new generating sectors of the last several years in the United States have been Renewable Energy and Natural Gas. In the predictions shown on the chart, these two sectors represent 83% of all new generation in the United States for the period 2008-2035.

World Wind Energy Potential

The estimates of available wind energy in the world range from 300 to 900 TW with 70% over open oceans. The lower estimate of 300 TW supplied over a year with an average production of the wind at 25% would produce 300 TW x 8,760 hrs/yr x .25 CF = 657,000 TWh annually. Using the lower estimate and limiting development to land sites would indicate a potential of 90 TW. This would provide 90 TW x 8,760 hrs/yr x 0.25 CF = 197,000 TWh annually; seven times the current total world consumption.

The German Advisory Council on Global Change (WBGU) in 2003 indicated the global energy potential from onshore and offshore wind developments was 278,000 TWh per year. They assumed a reasonable achievable target would be 39,000 TWh per year, more than double the current global electricity demand of 17,000 TWh (17 petawatt hours) in 2005.

Stanford University's Global Climate and Energy Project did an evaluation of the potential of wind power. They indicated from their findings that using only 20% of the potential could supply seven times the world's electricity demand in the year 2000. The demand in 2000 was 15,000 TWh. Seven times this would be 105,000 TWh, and if this was only 20% of the potential, the full potential would be 525,000 TWh.

Estimating the world wind energy potential is an immense challenge, and it is interesting to note the relative agreement of these three independent sources. *They clearly show that the reasonably available wind energy on a world scale is many times the total electric and other power requirements of humanity for the foreseeable future.*

Global Wind Energy Installations at the End of 2009

Worldwide installations of wind energy in 2009 totalled 38,300 MW*, a new record. The total installed wind energy at the end of 2009 was 158,500 MW. Ref. GWEC.

* This is 38,300 / 137,000 = 28% of the annual rate needed to meet the GWEC long-term target; see the projections following. China led all other countries with nearly 14,000 MW, 40% more than the next closest which was the United States.

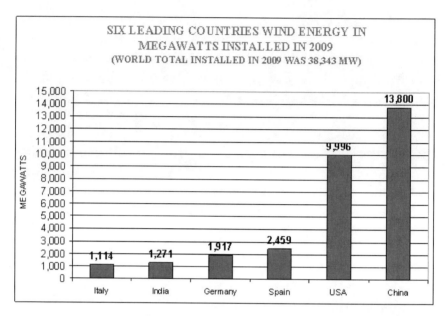

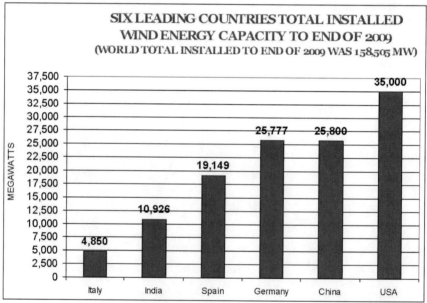

Wind Energy Installations End of 2010

China, for the second year, exceeded all other countries with wind energy installations of 16,000 MW, compared to Europe with 9,300 MW and the United States with 5,100 MW. Total wind energy installations at the end of 2010 were 84,000 MW for Europe; 42,000 MW for China; and 40,000 MW for the United States.

Wind Energy Global Projections by GWEC, GHG Offsets, and Investment Capital

The Global Wind Energy Council (GWEC) reports 158,500 MW (158.5 gigawatts) of total installed wind energy at the end of 2009 will produce 340 TWh (terrawatt hours) of electricity annually. They apply an annual capacity factor of 25%.

$$340 \times 10^{12} \text{ watts/yr} / 158.5 \times 10^9 \text{ watts} / 8,760 \text{ hrs/yr} = 0.25 \text{ capacity factor}$$

They claim this will save or offset 207 million tonnes of carbon dioxide (CO_2) (projected from GWEC 2008 data) annually. The offset savings per MWh are indicated to be

$$207 \times 10^6 \text{ tonnes/yr} / 340 \times 10^6 \text{ MWh /yr} = 0.61 \text{ tonnes / MWh}.$$

GWEC Target of 16.5% of World Electricity Use by 2020

The Global Wind Energy Council (GWEC) proposes 16.5% of the world's electrical power by 2020 and 34% by 2050 from wind energy.

The GWEC proposed plan for 2020 of 16.5% of the world power requirements would be

$$0.165 \times 25 \times 10^{15} \text{ Whrs} = 4.1 \times 10^{15} \text{ Whrs. (petawatt hours)}$$
(Ref. The 25 pettawatt hours from Asko Vuorinen report).

Converting this to wind power installations divided by 8,760 hour/year and by 0.25 capacity factor indicates the capacity required by 2020:

$$4.1 \times 10^{15} \text{ Whrs/yr} / 8,760 \text{ hrs/yr} / 0.25 \text{ CF} = 1,870 \times 10^9 \text{ W or 1,870 GW}.$$

The additional installed capacity over the eleven years would be 1,870 GW less the existing world installed capacity of 158 GW in 2009.

This would require an average of (1,870-158) GW / 11 years = 156 GW of new connected wind power annually for the eleven years from 2009 to 2020. *This is four times the total of installations made in 2009 of 38 GW.*

The average annual investment for 156 GW at current costs of $2 million per MW would be

> 156,000 MW x $2 million / MW = $312 billion world total annually.

GWEC Target of 34% by 2050

The GWEC target of 34% of the world's electricity requirements from wind energy by 2050 would be

> $0.34 \times 37 \times 10^{15}$ Whrs = 12.6×10^{15} Whrs (12.6 petawatt hours)

Note: World electricity consumption by 2050 per Asko Vuorinen, Optimal Power Systems, 2007, was 37,000 TWh from the lower of two scenarios.

In terms of the installed wind farm capacity required to meet this target:

> $(12.6-.158) \times 10^{15}$ Whrs/yr / 8760 hrs/yr / 0.25 CF = $5,750 \times 10^9$ W or 5,750 GW

The average annual target for 2050 would be 5,750 GW / (2050-2009) = 140 GW / yr. The capacity installed worldwide in 2009 was 38 GW, 27% of the annual target.

At a current average installed cost of $2 million per MW, the total investment required annually over the 41 years would be:

5,750,000 MW × $2,000,000 / MW / 41 years = $280 billion/yr

Applying the U.S. projection for total world electricity consumption of 45,500 billion kWh by 2050, the installed "wind energy" capacity would be 0.34 × 45,500 billion kWh/yr / 8,760 hrs/yr / 0.25 C.F. = 7,064 GW.

The additional installed capacity annually from 2009 to 2050 would be (7,064-0.158) GW / 41 years = 172 GW / yr.

At an average installed cost of $2 million per MW, the annual investment required would be $344 billion. Mass production may reduce the installation costs to $1.5 or $1.0 million per MW—$258 billion or $172 billion annually.

CANWEA in 2008 projected investment in new wind energy projects worldwide at $1 trillion by 2020. That would average $83 billion annually with an average of 42,000 MW of new installed capacity annually at an installation cost of $2 million per MW.

During the projected period of 42 years, the total world requirement for electricity is indicated to grow by over 20,000 TWh or 476 TWh/yr. This would demand an average energy production growth rate of 54 GW / yr.

$$20,000 \times 10^{12} \text{ Whrs} / 42 \text{ yrs.} = 476 \times 10^{12} \text{ Whrs/yr}$$
$$476 \times 10^{12} \text{ Whrs/yr} / 8760 \text{ hrs/yr} = 54 \times 10^{9} \text{ W or } 54 \text{ GW/yr}$$

If this demand was to be met by coal-fired power with an average demand of 38%, the total installed capacity would be 54 GW / 0.38 = 142 GW. If the new coal-fired power plants require carbon sequestration, the capital cost per MW would be in the order of $3,500,000. The annual capital investment over the period would be

54 GW / yr × 1,000 MW/GW/ 0.38 CF × $3.5 × 10^6 per MW = $497 billion/ yr.

If this total demand was to be met by natural gas-fired turbine generators and assuming there would be sufficient long-term supply, the average demand would be in the order of 40% (quicker response). This would require an installed capacity of 54 GW / 0.40 = 135 GW. At an average installed cost of $1 million per MW, the capital requirement would be $135 billion/yr.

Wind Energy Development in China

The International Energy Outlook 2009 (DOE-U.S.) expects wind energy in China to increase to 315 billion kilowatt hours in 2030. See the chart following.

With 20% of the world's population, China's consumption of electricity is only 10% of the world's total. Wind power costs worldwide have been reduced 80% per MW in the past twenty years, and wind energy is clearly a significant part of China's new energy policy.

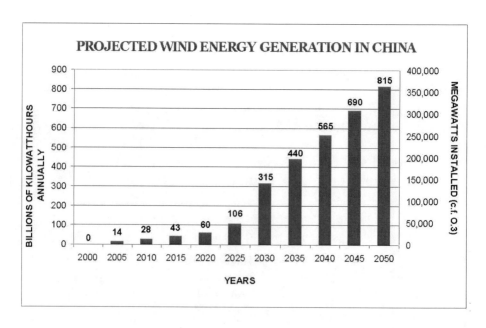

Note: From the International Energy Outlook 2009 of the EIA/DOE (U.S.) with 2035 to 2050 extrapolated at the average rate of increase from 2020 to 2030. Installed capacity in Megawatts is added, based on a capacity factor of 0.3. Some references indicated the average capacity factor experienced in China is 0.25.

China had established a target of 10,000 MW of installed wind energy by 2010. It achieved a cumulative total in 2009 of 25,200 MW. According to Wikipedia, China has a target of 100,000 MW by 2020, which would be 220 billion kWh/yr, almost four times the target indicated in the preceding chart.

Offshore Wind Energy Development in Europe

Eight new wind farms were established offshore in Europe in 2009 with a capacity of 577 MW. This is a growth rate of 54% over the 373 MW installed in 2008. The European Wind Energy Association (EWEA) is projecting an additional 1,000 MW of offshore wind farm capacity in 2010. Seventeen new offshore wind farms are under development in Europe with a potential capacity of 3,500 MW. Additionally there are 52 offshore wind farms which have consent for construction in European waters with a capacity of more than 16,000 MW, half of this in Germany.

The European year 2009 offshore wind farm investment has been indicated at E1.5 billion. This would be about $3.5 million per MW. This increased cost per MW compared to wind farms on land at about $2 million per MW is offset to a considerable degree by the higher capacity factor for offshore wind turbine generators.

Renewables in Europe

Thirty-nine percent of all new power generation in Europe in 2009 was wind power, followed by natural gas at 26% and solar photovoltaics at 16%. Overall, 61% of all new electrical power generation in Europe in 2009 was from renewables.

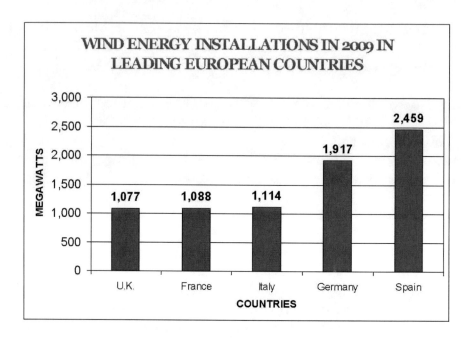

Onshore wind farms total of 7,070 MW in Europe in 2009 are reported to have an investment of E11.5 billion or $16 billion U.S. for an average of $2.26 million per MW installed cost.

Renewable Energy in Spain and the Hydrogen Storage Buffer

(Ref: University of Valencia)

Current wind power generation in Spain is 16,800 MW (2008) with planned upgrading to 20,000 MW by 2010. Average capacity factor is 22%.

This study predicts for a hydrogen buffer system to reduce variability of supply a volume of stored hydrogen in the order of 50 Nm3 per MW of installed wind power would be required (50 Nm3 / 11 Nm3/kg = 4.5 kg of hydrogen). At the indicated 38% efficiency of conversion from hydrogen by electrolysis to A.C. electricity through a fuel cell (see table $0.9 \times 0.42 = 0.38$), the electricity, which could be supplied from this "buffer" to the grid, would be

39.4 kWe per kg of H2 x 4.5 kg H_2 / MW x 0.38 effy = 67 kW/MW

For the installed wind farm capacity by 2008 of 16,800 MW, all buffered as indicated, the total electrical energy available to feed to the grid from fuel cells would be

67 kW/MW x 16,800 MW = 1,123,600 kW or 1,124 MW.

With an improved efficiency as projected in the University of Valencia report of 45% from 23%, the energy available from the hydrogen buffer would be

45/23 x 1,124 MW = 2,200 MW.

The report proposes electrolyser efficiency can be increased 10% over the current values for alkaline units. It indicates solid oxide fuel cells are expected to provide an efficiency increase of 20%, and a 5% increase in efficiency of the power electronics systems can be achieved.

The report predicts with a 45% efficiency of the buffer system and generating hydrogen for 6 hours during low-demand periods, the payout period would be 12 years. Higher buffer system efficiencies and higher electricity prices in high-demand hours would reduce this payout period.

Current and Targeted Efficiencies for Each Subsystem in the Hydrogen Buffer

System Efficiency %	Electrolyser Supply	Electrolyser	Fuel Cell	DC/AC Power Converter	Global %
Achieved	90	67	42	90	23
Target Objective	95	75	66	95	45

Ref: University of Valencia, Spain

Spanish Record

On November 8, 2009, wind energy in Spain broke a record. It supplied 53% of total electrical demand from 3 a.m. to 8:30 a.m. at 10,170 MW (ref. North American WindPower).

European Wind Energy Record

The European Wind Energy Council (EWEC) reported for 2009 installation of new wind energy facilities represented 39% of all new generation installed in 2009, with the second largest component of new 2009 generation at 26% from natural gas.

Renewable Energy Sources

The primary sources of renewable energy are biomass, solar, wind, stored hydro, run-of-river hydro, wave, tidal, and geothermal. Except for stored hydro, biomass, and geothermal, they are variable and dependent on nature.

Wind and biomass currently offer the most favorable costs of renewable energy sources after both hydro systems.

Wind energy has the fastest growth of the variables.

Solar energy has achieved major advances in the last decade. Utility scale generating facilities are operating and under construction.

Wind Energy Development in Canada

Samsung C&T Corporation and Korea Electric Power Corporation have announced plans to invest Can$7 billion to generate 2,500 MW of wind and solar power in Ontario with associated facilities. (Ref. North American Windpower, Jan. 22, 2010.)

United States Electricity Growth Rate

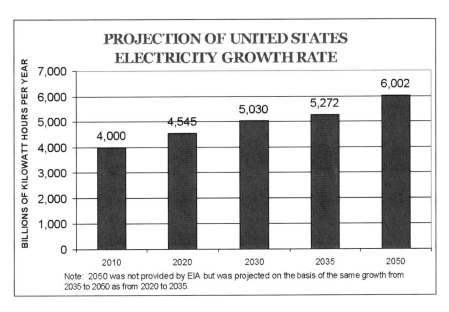

The DOE/EIA provided projections of U.S. electrical power usage in their Annual Energy Outlook issued December 14, 2009:

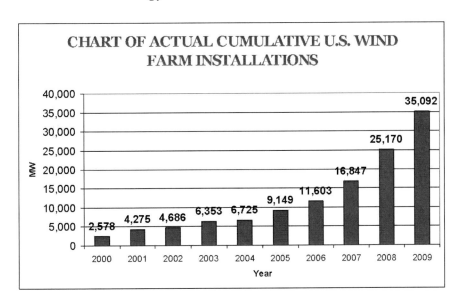

Cumulative Installed Wind Energy in the United States

In 2005 and in 2006, in the United States, wind farms provided its second largest source of new electricity supply, second only to natural gas. Installed capacity increased by 45% in 2007 with a 50% increase year over year in 2008 and by 40% in 2009. At the end of 2008, the United States had more installed wind energy than any other country, for the first time exceeding that of Germany. The estimated wind energy production in the United States in 2007 was 31 billion kWh.

Referring to the preceding chart, the growth in installed wind energy in the United States from 2004 to 2009 was 35,092-6,726 = 28,367 MW over the five years or an average of 5,700 MW per year.

AWEA reported growth of wind energy installations in 2009 of 9,922 MW in the United States, a new record, with the total to the end of 2009 at 35,090 MW.

Wind energy production in the United States for a full year with installed capacity at the end of 2009 of 37,092 MW would be

> 37,092 MW x 8,760 hrs/yr x 0.3 CF = 97 x 10^6 MWh or 97 billion kWh. (2.4 % of the U.S. total electricity usage projected for 2010).

10,000 MW of wind energy saves 600 million cubic feet of natural gas per day.

10,000 MW of wind energy saves 11 million tonnes of carbon dioxide emissions per year.

The wind energy potential of the top twenty American states could provide 1,200,000 MWh/hr (two-thirds of the U.S. total consumption in 2006). The U.S. total consumption in 2006 was

> 1,800,000 MWh/hr x 8,760 hrs/yr = 15.8 x 10^9 MWh/yr approx. 16,000 TWh/yr.

United States Wind Energy Potential

UNITED STATES WIND ENERGY POTENTIAL

THE TOP TWENTY STATES for wind energy potential, as measured by annual energy potential in billions of kWhs, factoring in environmental and land use exclusions for wind class of 5 and higher.

	Billions of kWhs	MW @ 0.3 CF		Billions of kWhs	MW @ 0.3 CF
North Dakota	1,210	460,000	Nebraska	868	330,000
Texas	1,190	452,000	Minnesota	857	326,000
Kansas	1,070	407,000	Wyoming	747	284,000
South Dakota	1,030	392,000	Oklahoma	725	276,000
Montana	1,020	388,000	Iowa	551	209,000
			Other Top Ten States	1,462	556,000
			TOTAL	10,470	3,980,000

At an average wind capacity factor of 30%, the total potential could provide 1,200,000 MWh/hour, and the connected wind energy capacity could be 4,000,000 MW.

The total potential wind energy resource of North America could be 200% of the U.S. potential or 2,400,000 MWh/hour.

United States Offshore Wind Energy Potential

The U.S. DOE's National Renewable Energy Laboratory in September, 2010 released a report which provides the estimated offshore wind energy potential for the United States of 4,150

GW, nameplate capacity basis. It covers offshore areas within 50 nautical miles of shore, an average annual wind speed of at least 7 meters per second (16 miles per hour) at a height of 295 feet. No deductions have been made for environmental or other considerations.

Wind Energy Growth Projections in the United States

As noted preceding, in the President of the United States State of the Union Address in 2006, it was proposed to produce 20% of the U.S. electric power requirements from wind energy. In 2008, the U.S. Department of Energy issued a report "20% Wind Energy by 2030: Increasing Wind Energy's Contribution to the U.S. Electricity Supply." (Ref. North American WindPower Feb. 2009.)

In 2005, the total U.S. electricity consumption was 3.816×10^{12} kWh (kilowatt hours), and in 2007, it was estimated to be 3.892×10^{12} kWh. This indicated an increase of 0.038×10^{12} kWh per year or 1%/yr. If this rate of growth in demand continues, then by 2030, the requirement will be 4.893×10^{12} kWh.

> A target of 20% from wind energy would require 4.893×10^{12} kWh $\times 0.20 = 0.979 \times 10^{12}$ kWh.

To determine the generating capacity in megawatts (MW) of the wind energy farms required in 2030 to meet this projection, it is necessary to convert the kWh to megawatt hours (MWh), convert from annual to hourly to obtain megawatts, and apply a capacity factor (CF) of 30% to provide for the average production of the variable resource.

> 0.979×10^{12} kWh/yr / 1,000 kWh / MWh / 8,760 hrs/yr / 0.30 CF = 373,000 MW.

The growth in wind energy in the United States, see the previous chart, was 40% annually from 2007 to 2009 and an average of 30% per year over the years 2005 through 2009. From 2009 through 2030, an average annual growth of 12.5% would achieve an installed capacity of 375,000 MW by 2031. At a capacity factor of 30%, this would supply 20% of the projected total electricity demand in the United States by that year. This growth is illustrated by the following chart.

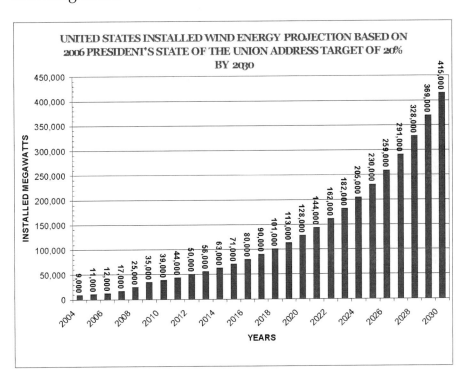

United States: World's Largest Wind Farm in Progress

This project in Oregon State will be 845 megawatts supplying electric power to Southern California Edison. It is reported by EERE (U.S. DOE) that the facility will provide avoidance of emissions of 1.2 million tons of carbon dioxide annually. Assuming a CF of 0.3, the annual megawatt-hour generation would be:

$$845 \text{ MW} \times 8{,}760 \text{ hrs/yr} \times 0.3 \text{ CF} = 2.22 \times 10^6 \text{ MWh/yr}$$

The CO_2 avoidance would be 0.54 tons per MWh.

North American Clean Energy (September/October 2008)

A projection was provided in the article "On Renewable Energy Technologies" by Jami Krynski that the United States would reach a cumulative installed wind capacity of nearly 49,000 MW by the end of 2015. This is only slightly below the projection shown for 2015 on the preceding chart.

Wind Energy Costs

Wind energy costs remain stable over the long term. There are no fuel price increases. Contracts at preagreed prices can be fixed over 20 years and longer periods without inflation adjustments except for CPI on the minor G and A expenses.

Power-generating facilities are built for long-term operation, forty years and more. Inflation of fuel costs for thermal power generation is shown to result in consumer costs significantly higher than from wind energy.

A Hatch Energy (Windsight 2008) report for power costs in Alberta projected the cost of coal generation with carbon capture at $100 to $140/MWh (2007 $'s) and nuclear power about the same. They project new wind, hydro, and CCGT generation in the range of $80 to $120 per MWh.

The Ontario government has guaranteed "feed-in" tariffs for renewables of solar, wind, water, and biomass—regardless of size—over a twenty-year period. Higher premiums will be provided for smaller projects. Homeowners with solar power systems will supply to the grid at more than eighty cents per kWh. Wind power projects will obtain prices between $135 and $190 per MWh, hydro $129 per MWh, and biomass $122 per MWh. The policy is enabled by the *Green Energy Act*.

Capital costs for wind farms are generally reported in a plus or minus 10% of $2,000,000 per MW of installed capacity.

Canadian Hydro Developers have installed 197.8 MW of wind energy on Wolfe Island, Ontario, at a capital cost reported (Canwea Issue 120) at $475,000,000. This is Can$2,400,000 per MW—close to the $2,000,000 U.S. average—adjusted for exchange rate at the time.

North American WindPower, January 2010, reported on a Virginia-based company Catch the Wind Ltd. which has completed a trial with Nebraska Public Power District of a system to better align the oncoming wind along with increased gust detection. The system is reported to have increased energy output of a wind turbine by an average of 12.3%, possibly as much as 18%. This would have a significant impact on reducing the cost of wind power.

The International Energy Agency (IEA) division, IEA Wind, has a task force focusing on the cost of wind energy systems as reported in 2009. The IEA Cost of Wind Energy Task 26 notes that due to increasing prices for copper and steel as well as demand the installed costs of wind systems, using the United States as an example, increased from 2002 through 2007 by 27% to approximately $1,710 per kW. (Note: Alternative generating systems would have experienced the same increased costs relative to commodities.)

Major cost elements for development of a wind farm are:

- Wind turbine generators, in the range of 60 to 70% of the total project cost Generally includes guaranteed performance for the first five years.
- Transportation
- Infrastructure including site preparation, foundations, service roads, electrical systems, and transmission

- Development costs, including feasibility studies with the wind resource assessment, negotiations for power purchase agreements, environmental permitting, land use rights, stakeholder relationships and agreements. This cost can be the most difficult to estimate and can be relatively high in regions where permitting regulations are onerous.
- Financing costs
- Internal costs, including all negotiations and fees for the preceding, prior to completion of financing

Wind Output Compared with Normal Electricity Demand

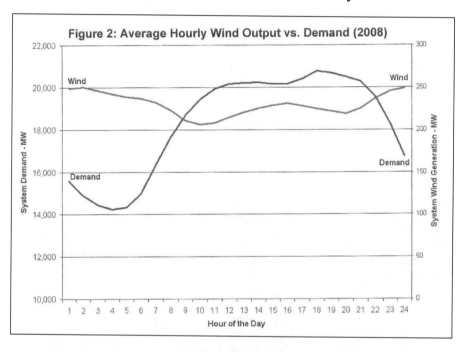

In many regions, the wind energy supply generally peaks during nighttime when the power demand is low. This chart above indicates this characteristic and supports the need to combine major wind energy systems with energy storage. The reference is the New York Independent System Operator report of March 2010.

California Energy Commission Renewables Prices

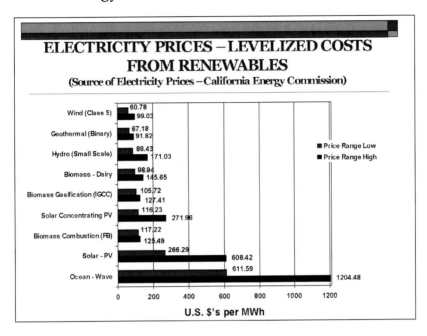

These costs show new installations of wind energy are lower than other forms of renewable electricity.

Quebec Hydro Energy Generation Payback Ratios

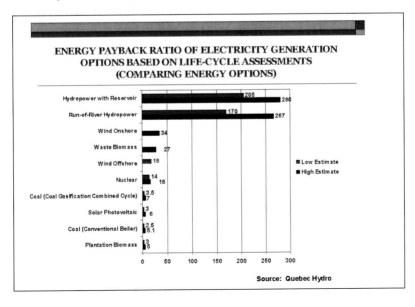

Puget Sound Energy Wind Energy Competitive

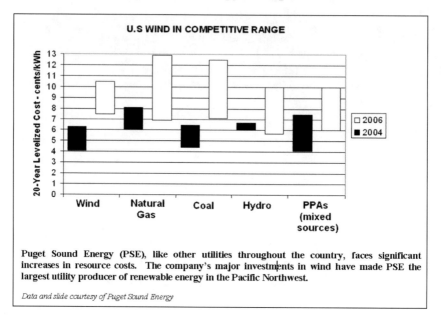

Puget Sound Energy (PSE), like other utilities throughout the country, faces significant increases in resource costs. The company's major investments in wind have made PSE the largest utility producer of renewable energy in the Pacific Northwest.

Data and slide courtesy of Puget Sound Energy

General References

Electricity Price Reform

The C.D. Howe Institute, quoted by CANWEA Jan. 26, 2010, recommends that electricity prices should fully link the consumer price to peak-period costs, environmental costs, and the high cost of new generation. It notes that most jurisdictions have relied on flat rates and price freezes for electricity, which may be politically expedient in the near term but have led to overconsumption, pollution, fiscal stress, and excess pressures on the generation systems.

Wind to Hydrogen Development Projects (W2H2)

An initial system for a pilot operation converting wind energy to hydrogen is developed by NREL and Xcel Energy Inc (a U.S. utility with head offices in Minneapolis). It is located near Boulder, Colorado.

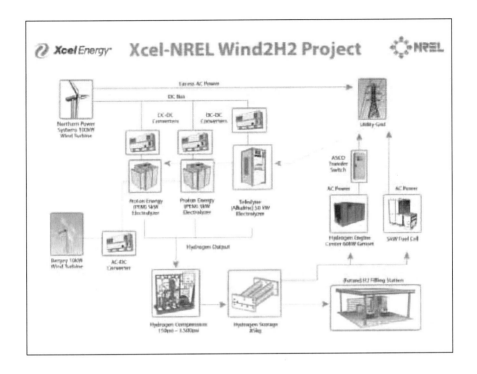

Danish W2H2CHP Energy Plant on Island of Lolland

This island community produces 50% more wind power than they use, and the excess is now (2007) being converted by electrolysis to hydrogen. The oxygen from electrolysis is used to speed up the biological process in the municipal water treatment plant.

The hydrogen at 6 bars fuels two PEM fuel cell combined heat and power (CHP) plants of 2 kW and 6.5 kW.

Thirty-five homes in the village of Vestenskov on the island will be installing micro CHP units to provide each with its own electricity and heat.

Pilot Wind to Hydrogen to Fuel Cells Plants

There are a number of these development small-scale plants in operation.

Potential of the World's Deserts

The deserts, both hot and cold climes, generally provide a significant wind resource. To exploit these will require relatively large development programs involving major transmission systems to reach the nearest load centers, together with means of storage to provide a degree of power on demand. The North African and Middle East deserts hold the potential to supply major quantities of renewable power to the load centers of Europe. A combination of wind and solar thermal may provide the best answer to power on demand. Solar thermal requires cooling water with the alternative, where water is lacking, of air condensers which would be very effective at nighttime.

Supply to Electric Vehicles

There is a current emphasis on placing millions of electric battery-driven vehicles in service. Generally these will be plugged in at night to recharge the batteries. The lowest-cost nighttime power is from coal-fired plants as their only variable cost, which would apply, is the coal. Coal is the lowest cost energy resource (before conversion to usable energy). This is also the highest producer of greenhouse gases, indicating that the electric battery vehicles will be major emitters of GHGs on a life-cycle basis even with no direct emissions as they operate.

Wind energy generally produces at its highest level during nighttime, when utilities normally have the least demand for it. Policies and incentives can readily direct increased demand for wind energy to supply the charging requirements of electric battery vehicles. This would provide these vehicles with truly emissions-free operation.

United States EPA and Coal-fired Plants

The *North American Windpower* issue of December 2010 reports on a Brattle Group study suggesting that EPA requirements for air quality and water could result in over 50,000 MW of coal-fired plants being retired.

Observations and Conclusions

- World electricity production is (17 / 138) = 12% of the total World Energy Consumption.
- This indicates that while replacing a significant portion of fossil fuel-powered electrical generation with renewable electric power will reduce large volumes of Green House Gases (GHGs), the other sources of energy will continue to be major contributors of GHGs, including motorized vehicles.
- Converting energy used in vehicles from gasoline and diesel to electric battery power will not significantly reduce GHGs so long as the electricity is from fossil-fueled generation.
- Wind energy, in addition to supplying the electric grid, converted to hydrogen used in fuel cell vehicles will significantly reduce GHG emissions.
- Wind energy growth projections are based only on supplying a portion of the electrical power requirements and do not include targets for supply to fuel-cell vehicles.
- Land requirements for wind farms take up only 3% of the total land area of the development.
- Projections of future growth of electrical power in the United States show renewable energy and natural gas fueled generation as the significant major sources.
- Of the three regions of the world leading in development of wind energy, Europe has achieved a high level of integration of wind energy and solar into its grid. It has been the world leader in providing policies and incentives for this investment. China, with one-half the total electricity production of the United States, is now second only to Europe in total installed wind energy.
- Wind energy installations can be financially viable in a much smaller scale than new fossil fuel and nuclear plants.
- Levelized unit cost of wind energy, with no direct cost for the resource and no inflation on it, is equal to or less than that of new carbon-sequestered fossil fuel and nuclear plants.

- Long-term supply of wind energy has no natural restriction whereas fossil fuel supply is finite, and much of it is subject to international risk and competition.
- For wind energy to become a significant and desirable source to replace foreign oil and reduce fossil-fueled power generation, the development of storage systems to reduce the impact of its variability and the conversion to hydrogen for fuel cell vehicles needs the encouragement of policies and some degree of incentives until a level of mass production—together with the continuing significant improvement in yield—is reached when installation costs become increasingly favorable.
- Development of wind energy with energy storage and distributed power centers will dramatically reduce the need for new transmission lines and new fossil-fueled power generation to meet peak demands.
- Incentives need only to be twenty percent or less of the installed cost of wind energy; private investment will provide the major capital requirement.

Chapter 4

RENEWABLE SOLAR ENERGY SOURCES & COSTS

Solar Power—Photovoltaic

Photons, particles of light, are absorbed by a layer of silicon, the top anode, on a photovoltaic (PV) panel. The photons knock electrons loose from atoms. Very fine crystalline silicone converts the sunlight into electrical charges. They move from the positive anode to the negative cathode which releases electrons that will flow through a connected load and back to the positive anode plate.

Edmond Becquerel, a French scientist, first discovered the photovoltaic effect in 1829. Albert Einstein wrote a paper explaining the photovoltaic effect in 1905. He won the Noel Prize in 1921. In the 1970s, Elliot Berman developed solar cells, reducing the cost from $100 per watt to $20 per watt.

Electricity from Solar Energy

The two main methods of solar electric power generation, mirrored sun rays and photovoltaic systems, have catapulted from the experimental stage to utility upgrade plants within the past five to ten years.

Costs are being reduced toward a competitive level with conventional power generation and other renewable energy sources.

Projections are being made by the industry that as much as 40% of the world's electricity can be supplied by solar systems.

As with other renewables, the resource offers many advantages, no loss of supply, no inflation, and no GHG emissions.

The most productive sites are desert areas so that considerable land areas required should not have to compete with agricultural land.

Solar energy's main obstacles are cost, lack of transmission lines—which are increasingly difficult to permit—and being able to supply the electricity on demand. Conversion to hydrogen energy for storage with return to the grid through fuel cells to meet the demand is a potential solution along with other energy storage methods described in Chapter 14.

The total available world solar energy resources are estimated to be 3.8 YJ/yr or 120,000 TW (average rate over a year). Less than 0.02% are considered sufficient to meet today's world energy demands. Installations of grid-connected photovoltaics increased by 83% in 2009 to an accumulated total of 15,000 MW. Nearly one-half of the increase was in Germany, followed by Japan. Ref. Wikipedia.

Solar Thermal Systems

Solar thermal uses reflectors to focus the sun's rays on a pipe or other heat accumulators. Some accumulators contain an organic fluid. The heated fluid is passed through an exchanger with water which is converted to steam. This steam then enters a turbine driving an electric generator. The system can be integrated with a thermal power plant using other energy sources to produce steam. The steam from both sources can be run through a common turbine generator. Steam from the thermal plant can be regulated to balance changes in supply of steam from the solar thermal facility. The solar thermal system obtains most of its energy from the infrared portion, the long wavelength, of the sun's rays. The solar thermal system, because of its dependence on additional processing facilities, does not lend itself to isolated applications as well as the photovoltaics, except for heating purposes. The solar thermal system does appear suited to applications in utilities with thermal power plants operating in regions where a percentage of renewable energy is mandated by government, due to the steam-generating synergies.

Solar Photovoltaic Systems

Solar Photovoltaic systems utilize the ultra-violet spectrum of the sun's rays, with photons, particles of light, striking the surface sensitive panels to release electrons. The conversion of the sun's energy is directly to electricity. The efficiency of conversion depends largely on the construction and type of mineral used for the panel surface. Efficiencies can range from a little over 10% to as high as 19% with recent innovations claiming close to 20%. Capacity factors vary, with an indication for California installations of production of 1,500 kWh/year for each installed watt. If this is applied against the number of hours in the year—8,760—then the capacity factor (CF) would be 17%. This compares to an average for most wind-power installations of 25%, with many over 30% and offshore installations exceeding 40%. Photovoltaic installations lend themselves to isolated off-grid applications, and if combined with electrolysers to convert excess electricity to hydrogen for reconversion to electricity through fuel cells, then a reasonably firm energy supply can be provided.

Technology

It has been reported that in certain manufacturing processes, such as placing the silicon film, a medium is used which results in some spillage. This is cleaned up with a compound including fluorine, which has an environmental impact about 23,000 times greater than carbon dioxide.

There are six major technologies commercially developed to date for production of energy from solar power.

- thin-film silicon (TF-Si)
- cadmium telluride (CdTe)
- copper indium gallium diselenide (CIGS)
- crystalline silico (c-Si)
- high-concentration photovoltaics (HCPV)
- concentrating solar power (CSP)

Efficiencies (Yields) of Photovoltaic Systems

Some examples from industry publications serve to indicate the range of efficiencies anticipated. Efficiency of the PV system is the actual kilowatt-hour produced annually divided by the sum of the kilowatt capacity rating of the installation and the number of hours in the year 8,760.

Solar Industry of October 27th, 2010, described a project by Cenergy Power with a 354 kW rooftop PV system at Del Mar Farms. They expect it to produce 660,000 KWh annually. The predicted efficiency would be

$$660{,}000 \text{ kWh} / (354 \text{ kW} \times 8{,}760 \text{ hrs/yr}) = 21\%.$$

Wikipedia report for solar power in Germany capacity installed in 2009 of 8,877 MW and generation of 6,200 GWh. The efficiency or capacity factor indicated

$$6{,}200 \times 10^9 \text{ Wh/yr} / (8.877 \times 10^9 \text{ W} \times 8{,}760 \text{ hrs/yr}) = .08$$
or 8% (yield)

Yields and Efficiencies of Photovoltaic Systems

The efficiencies of solar panels is dependent on the components used in the film, the method of manufacture, and other factors. Efficiency would be the quantity of electricity produced by a panel relative to the amount of energy received by sunlight.

Annual yield is the result of a number of factors:

- Panel efficiency—E
- Directly: Number of hours of sunlight annually—NHS
- Inversely: Degrees of latitude from the equator—$1 / L^0$
- Directly: Angular rotation of the panel—AR
- Inversely: Smog Intensity—$1 / SI$
- Deterioration—D

When projecting the kilowatt-hour output of a planned photovoltaic installation, it is necessary to evaluate all of these factors. The net effect of each is not readily estimated, and experience with similar or identical installations is the best measure.

Status of Solar Energy Installations:

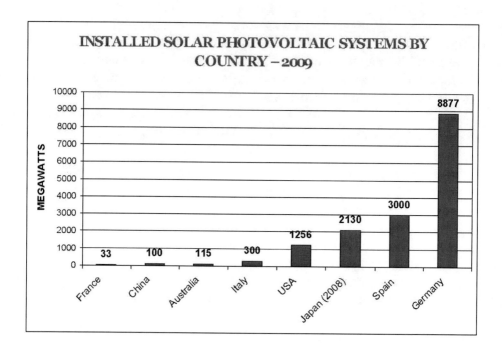

Germany is the most advanced country for solar photovoltaic energy installations, with 8,877 MW cumulative in 2009. The primary incentive is their feed-in tariff. The cost of this subsidy is one billion euros per month ($1.4 billion U.S.), paid by the rate payers. They produced 6,200 GWh in 2009, which was 1.1% of Germany's total electrical generation. A potential target for 2050 is 25% of the total power generated. Yield in 2009 was

$$6,200,000 \text{ MWh / yr / } (8,877 \text{ MW} \times 8,760 \text{ hrs/yr}) = .08$$
or 8%

IMS Research reports PV installations in 2010, worldwide, totaled 17,500 MW for a cumulative total of 37,500 MW. They project additional installations in 2011 of 20,500 MW—bringing the world total to 58,000 MW.

IMS estimates for 2011 there will be 22 countries installing over 50 MW, 18 countries installing over 100 MW, and 4 countries installing at least 1,000 MW. They expect, in 2015, there will be at least 34 countries installing over 100 MW each.

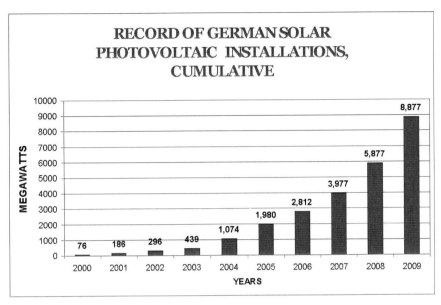

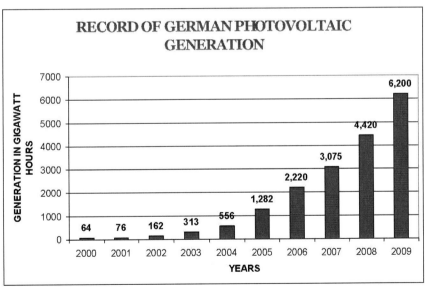

France had installed photovoltaic systems totaling 33 MW in 2006. The country has set a target of 3,000 MW by 2020. They have a target for wind energy from 810 MW in 2006 to 25,000 MW by 2020.

Spain in 2009 had 3,000 MW of photovoltaic installations, which supplied 2.8% of the electrical demand. Their largest photovoltaic system was 60 MW. Spain's PV installations grew from 470 MW in 2007 to 1,500 MW in 2008 to 3,000 MW in 2009. In 2008, Spain's industry minister forecasted expansion to 10,000 MW by 2020. However, there is some uncertainty due to their review of their FIT program.

Japan had a total installed PV connected to the grid of 1,900 MW in 2007 and added 230 MW in 2008. Their prime minister in 2008 set a target of 14,000 MW by 2020 and 53,000 MW by 2030. Japan, with just over 1% of the world's population, uses approximately 4% of the world's present energy consumption.

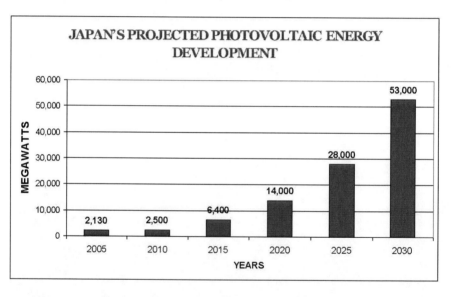

In the chart above, the installations of 2,130 MW in Japan was achieved in 2008.

Saudi Arabia, with desert area potential of 7,000 watts/square meter for 12 hours/day is planning a photovoltaic system to supply a water desalination plant.

Desertec, a German sponsored solar energy project, is proposed to bring Sahara desert solar energy by undersea high-voltage cable to Europe.

Projected Growth of Solar Energy in the United States

A projection of solar energy growth in the United States from about 1,000 MW in 2008 to 28,000 MW by 2016, subject to continuing government tax credits, was made by Navigant Consulting Inc. (NCI). They associated this with gross industry investments of $325 billion. They indicated an installed cost on the grid of $5.80 per Wpdc in 2008 projected to reduce to $4.00 per Wpdc in 2016 ($4 mm/MWpdc). At an average installed cost of $4.90 per Wpdc, the direct investment through 2016 would be $4.90/Wpdc x 27,000 MWpdc = $132 billion.

Note: Wpdc is a watt from photovoltaic (p) in direct current (dc). An MWpdc is one million watts.

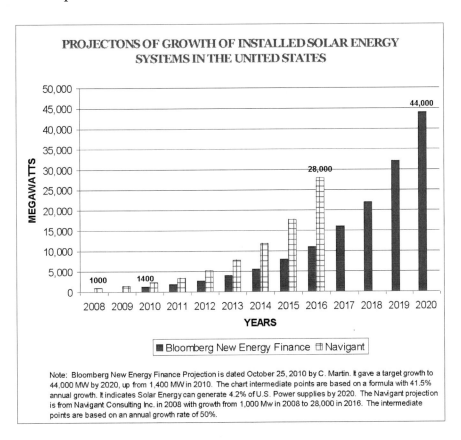

Note: Bloomberg New Energy Finance Projection is dated October 25, 2010 by C. Martin. It gave a target growth to 44,000 MW by 2020, up from 1,400 MW in 2010. The chart intermediate points are based on a formula with 41.5% annual growth. It indicates Solar Energy can generate 4.2% of U.S. Power supplies by 2020. The Navigant projection is from Navigant Consulting Inc. in 2008 with growth from 1,000 Mw in 2008 to 28,000 in 2016. The intermediate points are based on an annual growth rate of 50%.

Bloomberg New Energy projection suggests that solar power may meet 4.2% of U.S. electricity supplies by 2020. The U.S. electricity consumption is projected for 2020 at 4,545 x 10^9 kWh. The solar power component would be

4,545 x 10^9 kWh /yr x 0.042 = 191 x 10^9 kWh/yr.

Assuming a 10% yield (CF), this quantity of generation would require installed megawatts of

191 x10^9 kWh/yr / (1,000 kWh/MWh / 8,760 hrs/yr / 0.10 CF) = 22,000 MW

This is one-half the projection on the chart for 2020.

Observations of Solar Energy Development

A long-term common use, by almost everyone, of the solar cell is in handheld calculators which require no battery.

Photovoltaic (PV) conversion efficiency in commercial applications ranges from 7 to 17% of the energy in the sunlight to electricity potential. Some experimental PV cells are able to convert 40% of the sunlight to electricity potential, which indicates promise of the future.

The life cycle of modern solar panels is twenty to thirty years with minimum maintenance requirements.

A 10% efficiency PV system will generate up to 180 kWh per M^2.

A 1 kW system generates up to 1,800 kWh/year depending on its geographical location. The installed cost at $4 / Wpdc would be $4,000 per kW. Operating and maintenance costs (O and M), per NCI, would be in the order of $10 / kW/year.

Growth of photovoltaic generated power in Europe caused demand for the essential element silicon (processed) to become greater than the supply, raising the price of silicon from $200 per kilogram (kg) to over $450 per kg. This pushed up the price of solar PV panels from $3.25/Wpdc to $5/Wpdc. Expansion of production of silicon reduced its price and dropped the panel price to $3/Wpdc.

Rooftop solar systems act as insulators, reducing both heat and cold penetration into the buildings, saving energy demands and costs.

Solar thermal systems using evacuated tubes are claimed to take up only one-tenth the space of PV systems for equivalent energy with efficiencies up to 75%. (PV system efficiencies generally do not exceed 20%). These systems may become more commonly used as a renewable energy source, where heating and cooling energy requirements may exceed direct electrical energy requirements.

A development expected to reduce solar energy costs is high-concentration photovoltaics (HCPV). Concentrating the light will reduce the amount of PV material required. HCPV systems developed to date are achieving efficiencies of 35% and higher compared to flat-panel systems with 15% up to a maximum of 20%.

Solar Industry magazine reports a subsidiary of Boeing Co., Spectrolab, has started mass production of a solar cell referred to as C3MJ+. It is a concentrator photovoltaic (CPV) cell with a reported average efficiency of over 39%.

Water supply for solar PV and solar thermal generating plants is a concern. Both types of electric power generation have great potential in hot desert areas where, paradoxically, water sources are critically short. Washing PV panels is a water use. An alternative for solar thermal plants, which require cooling towers utilizing large quantities of water, is the air condenser. This does not function well during the heat of daytime but is very effective during the cool nighttime on the desert. And water is still a requirement in the heat-generation cycle. Another alternative is an Israeli company's development of a turbine-generator closed system using an organic fluid.

Solar Power Developments

These projects demonstrate the proven and successful technology of solar energy on a commercial basis when combined with appropriate policies and incentives. As such participatory encouragement continues and the quantity of generation grows, the resulting mass production effect will continue to result in dramatically reduced costs.

United States

Since 1991, the largest solar thermal power plant has been the 354 MW *Solar Energy Generating System, in the Mojave Desert of California,* which uses parabolic trough collectors.

MMA Renewable Ventures (Municipal Mortgage & Equity LLC) operates 35 MW of solar projects in the United States with 400 MW under development. These include 18 MW at Nellis Air Force Base in Nevada, which is the largest photovoltaic installation in the United States, a solar array at Denver International Airport and the world's largest thin-film solar installation using copper indium gallium diselenide technology. *MMA was purchased by Fotowatio,* a Spain-based independent solar power producer. (Ref. Solar Industry, April, 2009)

Sunpower with Iberdrola (Spain) are developing a 30 MW photovoltaic solar power plant in the San Luis Valley, Alamosa County, Colorado.

Lakeland Electric of Florida is installing a 24 MW photovoltaic system.

Florida Power and Light Co. is developing a 75 MW solar energy center to be the world's first hybrid solar energy plant. The solar system will produce steam to be used in conjunction with the steam from the heat recovery steam generator (HRSG) of a combined cycle natural gas turbine (CCGT) plant. The solar system will use 180,000 mirrors over 500 acres of land.

In New Mexico, Southwestern Public Service Co (Xcel), with SunEdison, have a 54 MW photovoltaic solar plant under construction in Lea and Eddy counties, utilizing over 400,000 panels in a 720-acre area.

Davidson County, North Carolina has a 16 MW solar farm under development by *SunEdison*, a subsidiary of MEMC Electronic Materials. It is predicted that the first 4 MW stage, with 14,000 panels, will generate over six million kilowatt-hours in the first year, for a 17% efficiency. Ref. Solar Industry, Feb., 2010.

The *Desoto Next Generation Solar Energy Center* installed, in 2009, a 25 MW photovoltaic facility. It will use 90,000 photovoltaic panels on 180 acres of land, providing electricity to 3,000 homes. The indicated land use is 7 acres per MW. The capacity factor (CF) can be calculated as follows:

> 3,000 homes x 1 kW / home / 8,760 hrs/yr = 26,280,000 kWh / yr
> 25 MW (installed) x 8,760 hrs / yr x 100% CF = 219,000 MWh / yr
> Actual capacity factor indicated will be
> 26,280 MWh / 219,000 MWh = 0.8 CF or 8% efficiency.

Note: The consumption per home used here in calculating the CF may be less than that expected by DeSota. If the home consumption is in the order of 1,500 kW, then the CF would be 0.12 and efficiency 12%.

Spanish renewable energy company *GA-Solar* and its parent company, *Corporacion Gestamp*, are planning a $1 billion photovoltaic project in *Guadalupe County, New Mexico*. It will utilize 2,500 acres of land, approximately 8 acres per MW, with a target of 300 MW, over a four-year period. The installed cost per MW is indicated to be $3.3 million. (Ref. Gov. Bill Richardson's office—Solar Industry Magazine.)

Peco, Philadelphia, has committed to purchase 6 MW of solar renewable energy credits (SREC's) in support of *Pennslvania's*

Alternative Energy Portfolio Standards, at an average price of just over $250 per megawatt-hour.

The *Arizona Public Service* have approval from the Arizona Corporation Commission to invest up to $500 million for 100 MW of photovoltaic power plants across Arizona, all by developers. The indicated installed cost is $5 million per MW. Ref: Solar Industry March, 2010.

A 5 MW multitechnology solar project at *Arizona Western College in Yuma,* Arizona, is being developed with five solar technologies—high-concentration solar, low-concentration solar, thin-film PV, monocrystalline silicon PV, and polycrystalline silicon—by PPA Partners Inc. of Morgan Hill California.

NRG Solar is proceeding with a 290 MW thin-film photovoltaic generating plant, Caliente Solaris, in Yuma County, Arizona. It will avoid the alternative generation with 237,000 tonnes of GHGs.

Pacific Gas and Electric are now receiving power from Sempra Generation at its 48 MW photovoltaic solar power plant near Boulder City, Nevada.

California

The *U.S. Bureau of Land Management* has received land use requests in *California* for over thirty solar thermal power projects for a total of 24,000 MW.

The California Energy commission reports two new photovoltaic projects—*550 MW at Carissa Plains and 250 MW in San Luis Obispo County*—to be supplied to *Pacific Gas and Electric* with start-ups in 2010 and 2011.

In 2008 a 2.3 MW solar photovoltaic system was installed on the roof of *Toyota's Parts Center in Ontario, California.*

Cogentrix Solar Services (Cogentrix Energy LLC) owns and operates two utility scale solar trough plants originally developed by Sunray Energy Inc. in 1984-1985 in San Bernadino County. They have *a capacity of 43 MW, supplying to Southern California Edison.*

Southern California Edison installed a 2.4 MW solar installation on 600,000 square feet of roof space at the Kaiser Distribution Center in Fontana, California. (Ref. Solar Industry, January, 2009). This is a good example of an installation close to a load demand center (a distributed power centre, or DPC) as it reduces the need for increased capacity of transmission and distribution facilities and it improves power availability in case of grid failures.

The *City of Los Angeles* has a project identified as Solar LA to provide 1.3 GW by 2020. This will consist of several facilities providing thermal and electrical energy.

Constellation Wines Gonzales Winery in Monterey County, California, will install a 1.2 MW Mitsubishi high-efficiency solar panel system on its 170,000-square-foot warehouse roof. It is projected to generate 1,175 kW of DC power.

A 300 MW photovoltaic facility is being developed in Southern California, near *Desert Center, to supply to Pacific Gas and Electric Co. (Ref. Solar Energy).*

Los Angeles Southwest College has 4 MW of panels on carport roofs and 2 MW of other systems, designed and constructed by *Chevron Energy Solutions*. It is projected to supply 5,000,000 kWh of electricity annually. The indicated capacity factor or efficiency is

>6 MW x 8,760 hrs/yr = 52,560,000 kWh / yr at 100%.
>5,000,000 kWh / yr / 52,560,000 kWh / yr x 100% = 10% plus.

Riverside, California, has a plant operated by the *Fresh and Easy* retail chain at which they installed 2 MW of photovoltaic solar energy facilities, producing 2.6 million kWh annually. This is an average of 300 kW/hr, indicating a capacity factor (or efficiency) of about 14%.

Hawthorne Machinery in California has a 155 kW photovoltaic facility which cost $1,275,000, about $8,000 per kW ($8 million per MW). A tracking system claims a 30% greater yield than a fixed system.

SunPower Corp is developing a solar photovoltaic 200 MW system, mainly on warehouse roofs to supply to *Southern California Edison*.

Prologis is installing a rooftop solar project of 11 MW in Southern California, as part of an agreement with Southern California Edison to provide a total of 100 MW. (Solar Industry May, 2010)

Southern California Edison is reported to have signed contracts for 260 MW of photovoltaic solar power projects, primarily with Silverado Power.

Bright Source Energy Inc. is developing a 392 MW solar electric generating system at Ivanpah, California. The plant will be known as ISEGS. NRG Energy is a partner. (Solar Industry).

Italy

A 72 MW photovoltaic power plant is being developed near the town of Rovigo in northeastern Italy by SunEdison (MEMC Electronic Materials Inc.). A *Solar Industry* magazine report indicates the project connected to the grid within a nine-month construction period.

Portugal

Amper Central Solar SA, owned by Acciona SA and 34% by Mitsubishi Corp., developed and operates the world's largest solar photovoltaic project in Amareleja, Portugal, at a capital cost of 261 million euros. It commenced operations in December 2008 with 45.8 MW and is projected to produce 93 million kWh of electricity annually.

The efficiency is indicated at 93,000 MWh over 8,760 hours, for an average of 10.62 MW, with an installed capacity of 45.8 MW to be a very favorable 23.2%.

The capital cost of about $280 million indicates a cost per installed MW of $6,000.

A 116 MW photovoltaic plant is under development in southern Portugal.

Slovakia

To encourage green energy, Slovakia has a feed-in tariff system with guaranteed grid connection, with a rate of 57 cents per kilowatt-hour for photovoltaic power ($570.00 per MWh).

Spain

The city of Arnedo in Spain has installed Spain's largest photovoltaic power-generating system.

A 50 MW concentrated solar power (CSP) plant is in development by Aries Ingenieria y Sistemas and Elecnor Group, with an indicated cost of 275 million euros. This appears to be in the order of $8 million per megawatt.

The 20 MW Beneixama photovoltaic was the world's largest when it was installed. Ref. Wikipedia.

Japan

Tokyo has developed a program for installation of photovoltaic systems on residences and apartments with a target of 1,000 MW.

Vatican

The Vatican is reported by Bloomberg (Van. Sun, April 18, 2009) to be building Europe's largest solar plant just north of Rome on a 300-hectare field. Plant capacity is projected at 100 MW with a cost of $800 million, $8 million per MW. It is expected most of the energy will be utilized as electricity with some solar heating applications. It is predicted that the plant will supply 40,000 households.

India

Astonfield Renewable Resources has an agreement to install 5 MW of solar power in Osiyan, Rajasthan, India, under the Jawaharlal Nehru national Sola Mission. (*Solar Industry* magazine May 12, 2010)

India's renewable energy secretary, Deepak Gupta, has indicated India will have a total installed photovoltaic generation of 1,100 MW by 2013—with a target of 20,000 MW—by 2022.

Taiwan

A Bloomberg report indicates Taiwan, which has a feed-in tariffs program for solar and wind developments, may reduce its tariff for solar projects from the present U.S. dollar equivalent of 0.37 cents per kilowatt-hour due to lower new plant costs.

World Solar Market

Generating component supply for solar energy in 2010 is projected to be 9,300 MW at an investment of $39 billion ($4.2 million per MW for the generating components). A projection is made of investment in the generating components of $77 billion for 26,400 MW in 2015 (costs reduced to $2.9 million per MW). Ref. Solar Industry Magazine March 10, 2010, and Lux Research.)

Observations and Conclusions

- Significantly large solar energy systems have been installed in the developed and developing regions, proving the long-term feasibility of solar power.
- Growth of solar energy as a measurable portion of total electricity consumption has been achieved.

- Only a fraction of the financially developable solar energy is needed to supply all of the world's energy needs.
- The technology of solar photovoltaic and thermal has advanced rapidly, is well proven and accepted.
- Photovoltaic and solar thermal systems have made major advances in capacity factor (yield) and in reducing costs.
- Research continues at a high level, supported by government and by businesses which are commercially successful in this field.
- Three factors which drive the success of solar energy are the need to reduce dependence on overseas crude oil, the need to develop this renewable source to replace the finite supply of fossil fuels, and the need to reduce greenhouse gas emissions from fossil-fueled generating sources which it replaces.
- Evaluating the cost of solar energy supplied to the grid must consider at least three factors when comparing it with fossil fuel sources.

    (1) Negligible inflation for solar energy
    (2) Virtually no greenhouse gasses which earns a carbon credit
    (3) No increase in energy source cost compared with the inevitable major rise in crude oil cost due to finite supply, growing demand, and politically unstable sources.

- Solar energy, with electrolysis, can supply all the world's vehicles with true emission-free operation.
- Solar photovoltaic direct current power, most profusely available in desert areas close to the oceans, can be used in desalinization plants to provide significant supplies of fresh water.
- Government policies and incentives are driving costs of solar energy down as mass production is achieved.

Chapter 5

HYDROGEN BY ELECTROLYSIS FROM RENEWABLES WITH COSTS

Mighty oaks from little acorns grow.

Hydrogen from Variable Renewables: Wind, Solar, Tidal, Wave, and Non-stored Hydro

Renewables of wind, solar, geothermal, tide, wave, and bioenergy (excluding hydro) are all expected to increase as a percentage of world power generation from 2.5% in 2007 to 8.6% in 2030. This would be the fastest growth rate of any power generation sector. (World Energy Outlook, International Energy Agency.)

"To improve the sustainability of the energy system, the European Union, together with other countries around the world, has embarked on the development of a hydrogen-including economy using hydrogen as an energy carrier. Unlike electricity, it can be stored in small or large quantities for long periods without significant losses. It can be produced from a wide variety of resources, including renewables, nuclear energy, and fossil fuels." Ref: Institute for Energy and Environment, Leipzig, Germany and European Commission, Institute for Energy, Petten, The Netherlands.

Variable renewable electric energy can be converted by electrolysis of water to hydrogen, a form of energy storage. The variability of the renewables becomes a firm supply to be used to meet demand.

It can be used to directly fuel vehicles, for refining oil to gasoline or diesel, for production of fertilizers, and for reconversion to peak-demand electricity, all with almost nil greenhouse gas emissions. It can also be used in the food industry, special chemical processes, protective atmosphere for some manufacturing, and power plant generator cooling.

The hydrogen can be transported by pipeline, with cavern storage en route (similar to systems used for many decades for storage of natural gas to take care of demand surges on pipelines). This maximizes the transmission facility utilization, whereas variable renewable electricity utilizes transmission only at one-third of its capacity.

There are many electrolyser manufacturers with units varying in efficiency, from 65% to claims over 90%, and costs from $5,000 per kilowatt to claims of $400 and lower. The United States DOE target price for electrolysers is $300 per kilowatt. General Electric's research center at Niskoyne, New York, has developed an electrolyzer suitable for mass production which could reduce the cost per kW close to the United States D.O.E. target.

There are electrolysers being developed to operate at high pressure thus reducing the cost of compression. High pressure for the hydrogen is necessary in order to store a sufficient volume, in the case of vehicles, to offer a reasonable travel between refuelings.

The two most common types of electrolysers are described as alkaline and proton-exchange-membrane (PEM). The alkaline type uses a 25% solution of potassium hydroxide (KOH). It is manufactured in capacities up to nearly 500 Nm3/hr (cubic metres per hour at normal pressure and temperature). The PEM type is limited by manufacture currently to a considerably lower maximum capacity; however it is more flexible in its ability to operate over a large range. It also lends itself to being able to produce hydrogen under pressure, eliminating the need for compressors with their energy consumption.

"Wind energy can be harnessed to provide electricity at some of the lowest costs available for new generation. Coupling wind turbines with hydrogen-generating electrolysers has the potential to provide low-cost, environmentally friendly electricity and hydrogen. In this way hydrogen generation can be a pathway for using wind energy to contribute directly to reducing the Nation's (U.S.) reliance on imported fuels." Ref: J.I. Levene, NREL, May, 2005.

U.S. DOE (Department of Energy) Electrolyser Development Goals

- Small facilities—250 kW, large facilities—3 MW.
- System efficiency 74% LHV (lower heat value) at 400 psi.

- Capital cost of $300/kWe (kilowatt equivalent) installed—large system.
- Hydrogen production cost of $2.85/kg hydrogen. The NREL target (May 25, 2004) from renewable integrated electrolytic hydrogen production was $2.50/kg.

General Observations

Creative financial arrangements are required with off-peak generators. This refers to drawing electricity from generating plants such as coal-fired units during nighttime off-peak hours when they are normally operating at only a small fraction of full load and their only expense for this incremental power is the cost of the coal.

Real-time optimization is needed for electricity and hydrogen production on the grid.

20% of light-duty vehicles require 12 million tons of hydrogen. At 74% LHV, there are 450 TWh of electricity required annually.

Electrolysis offers a pathway to carbon-free fuel for the transportation sector.

National Renewable Energy Lab (May 25, 2004) Wind-Hydrogen Pilot System

This proposed system maximizes the direct supply of wind energy to the grid with no loss of efficiency. When electricity from wind exceeds the grid demand, the surplus is directed to electrolysers, with the hydrogen used for vehicle fuel and for return to the grid through fuel cells during peak-demand periods. Surplus electricity from the grid during low-demand periods (low-value electricity) is supplied to electrolysers for hydrogen fueling vehicles. This hydrogen could also be returned to the grid through fuel cells during peak-demand periods. The electricity returned to the grid from hydrogen storage during peak demand also contributes value by being able to produce within the load centers and thus reduces not only additional generation for peak demand, but reduces demand during peaks on the transmission and distribution systems. It also provides backup in case of blackouts for emergency.

Target Cost/Performance for Class 6 Wind Resources

(NREL May 25, 2004)

Year	Installed Wind Farm Capital Cost $/kW	Capacity Factor Percent
2000	942	0.4
2010	754	0.5
2020	706	0.54

Energy Efficiency of Electrolysers

This is defined as the higher heating value of hydrogen produced by the electrolyser (HHV) divided by the amount of electrical energy consumed by the electrolysis system, not including ancillary equipment such as the converter, pump, etc. The energy required to separate water into hydrogen and oxygen is the same as the energy released when hydrogen and oxygen are combined to form liquid water. This HHV of hydrogen is 39 kWh/kg. This is the value used to calculate efficiency in this study as it is based on use of liquid water. The lower heating value of hydrogen (LHV) is the energy involved in formation of steam, 33 kWh/kg of hydrogen. This is not relevant in this study.

Proton Energy Systems

Ongoing research by Proton was reported at the FCHE conference, Feb., 2011, including reduced hydrogen permeation in a hydrocarbon membrane vs. a Nafion baseline, reduction in anode activation polarization of 80-100 mV with new catalyst formulations and 55% reduction in catalyst usage. They expect their research success will result in a gain of 15% in efficiency and a reduction of cell cost of over 35%.

Norsk Hydro Electrolysers

Norsk Hydro (now merged into Statoil-Norway), a major developer of electrolysers, report an electrode design with their specially developed catalytic coating which cuts cell voltage, reducing energy consumption by up to 20% compared with competitors. They note that 80%-90% of electrolyser operating cost is the electricity input. They claim a low 4.3 kWh/Nm^3 electricity-to-hydrogen-conversion at a pressure of 440 psig.

Note 1. 4.3 kWh/Nm^3 x 11 Nm^3/kg = 47.3 kWh/kg. This indicates an efficiency of 39/47.3 = 82.5%.

Their largest electrolyser (2007) generates 485 Nm^3 of hydrogen per hour (44 kg/hr) with 242.5 Nm^3 of oxygen, with skid frame floor space of 4 x 13.5 m. They can supply multiple units, to operate unattended.

NREL, J.I. Levene, May 2005, reports the cost of Norsk electrolysers at $800/kW. This would indicate a capital cost for an equivalent 100 kg/hr "bank" at 100 kg/hr x 47.3 kWh/kg x $800 = $3,800.

Norsk report their electrolysers provide automatic, unattended, continuous operation supplied with PLC systems.

Norsk electrolysers use a 25% KOH aqueous solution with feed water consumption of 1 liter per Nm^3 or 11 liters per kg of hydrogen.

Norsk observe that hydrogen is potentially the energy carrier of the future through water electrolysis.

They have developed and supplied remotely monitored and operated hydrogen by electrolysis production and fueling stations for cars and buses in a number of cities. In April 2003, they provided the first public hydrogen fueling station in the world, supplying buses. Two stations in Berlin were integrated, with regular gasoline-dispensing stations supplying hydrogen to 100 cars per day.

Powertech, B.C. Hydro, Off-grid Hydrogen

Powertech, a division of British Columbia Hydro, is installing a Hydrogen Assisted Renewable Power System at Bella Coola on the British Columbia coast. The town is isolated from the grid. It has a run-of-river small hydro generating station at Clayton Falls, which generates excess power at low-demand periods. This excess power will be converted by electrolysis to hydrogen. During high-demand periods, the hydrogen will be reconverted to electricity through fuel cells or ICEs.

Newfoundland and Labrador Hydro—H_2W_2D at Ramea

Newfoundland and Labrador Hydro is developing a wind-hydrogen-diesel energy system on the remote island of Ramea. Excess wind energy from wind turbine generators will be supplied to electrolysers producing hydrogen which is stored to provide electricity through fuel cells or ICEs when demand peaks and also to allow for shutdown of diesels when demand is low.

Xcel Energy, DOE/NREL, W_2H_2 Test Facility at Boulder

Xcel Energy (Minneapolis) and the U.S. Department of Energy's (DOE) National Renewable Energy Lab (NREL) developed a wind-to-hydrogen (W_2H_2) test facility at Boulder, Colorado. Wind energy is supplied through two electrolysers of different technology to produce hydrogen. One electrolyser is the Proton Energy HOGEN 40RE (PEM) unit and the other is Teledyne Energy System Inc. HM100 alkaline electrolyser. The plant cost was $2,000,000. The PEM unit is described as a Proton Exchange Membrane or a Polymer Electrolyte Membrane. The facility includes a 100 kW wind turbine, two 5 kW PEM electrolysers, one 50 kW alkaline electrolyser, and a 60 kW internal combustion engine (ICE).

Xcel Energy and Univ. of Minnesota W_2H_2

Xcel Energy and the University of Minnesota West Central Research are developing a W2H2 demonstration facility to supply hydrogen for production of anhydrous ammonia and for vehicles.

Sacramento, California SMUD SPH_2 Fueling Station

Sacramento Municipal Utility District (SMUD), California, has installed and is operating a solar-powered hydrogen (by electrolysis) fueling station for fuel-cell vehicles. The sponsors are SMUD, BP, Ford and the DOE. The solar array also delivers 80 kW of electricity to the grid.

Honda, Torrance, California SPH_2 R & D

Honda, since 2001, has operated a solar-powered water-electrolyser hydrogen station at its R and D center in Torrance, California. It utilizes thin film solar panels developed by Honda Engineering. A high—pressure production efficiency of 52 to 66% has been achieved.

ITM Power Home Hydrogen Electrolyser

ITM have developed an electrolyser for homes to produce hydrogen, from off-peak electricity, which is used through fuel cells to produce electricity, for home heating and for vehicle fueling.

Hydrogen Costs by Electrolysis

The NREL in Golden, Colorado, reports hydrogen produced from electrolysis at a wind farm of $5.70 per kilogram. This cost included power conversion, electrolyser, balance of plant, electricity input, operations, and maintenance. They gave a target by 2010 of grid connected at $2.85/kg and by 2015 of $2.75/kg from renewables (wind).

NREL, Xcel Energy, Basin Electric, Fort Collins Utilities, University of Minnesota, University of North Dakota

In 2006, NREL—with Xcel Energy, Basin Electric, Fort Collins Utilities, University of Minnesota, and the University of North Dakota—collaborated on a techno-economic analysis of central production of hydrogen from wind (hydrogen produced at the wind site and delivered to the point of use.)

Solid Oxide Electrolyser Cell Technology (SOEC)

Water electrolysis with a new SOEC resulted in a record-breaking current density of—3.6 A/cm^2 at a cell voltage of 1.48 V. There is a potential of high electricity efficiency.

Assuming surplus electricity to cost $3.60/GJ (1.30 cents per kWh) the hydrogen production is estimated to cost $5.00 U.S./GJ: equivalent to $30.00 U.S. per barrel of crude oil or $0.60/gge (gallon of gasoline equivalent). This pricing was based on hydrogen to electricity efficiency of 95%, with investment depreciation at 18% of the total production cost.

It is indicated that SOEC technology has a promising potential for production of hydrogen from renewable energy sources. However more R and D is required. Stability of the SOEC requires focus to ensure sufficient long life. The tested cells have already proven high stability used as SOFC's.

(Ref: Fuel Cells and solid State Chemistry Dept., Riso National Lab, Roskilde, Denmark; International Journal of Hydrogen Energy 32 (2007).

(Ref: Appendix for GJ units and conversions.)

Observations and Conclusions

Electrolysers for production of hydrogen have well-proven technology in commercial use over a hundred years, used extensively early in the 1900s to produce hydrogen for ammonia fertilizer from excess hydroelectricity (Norway and Canada).

Methods have been developed recently to mass-produce electrolysers with a dramatic reduction in the unit cost compared with the high proportion of hand labor for current models. As demand increases, the higher volumes will result in lower cost.

Banks of electrolysers can be installed at generating sources, such as wind farms or at load centers, remotely operated with no attendants on site.

Cost of hydrogen from renewables will be reduced as this means of storing and utilizing hydrogen expands with mass production of the electrolysers and associated components. A competitive comparison for use as a vehicle fuel follows:

COST OF HYDROGEN FROM RENEWABLES COMPARED
with
CRUDE OIL / GASOLINE AND METHANE / HYDROGEN FUELS

Electricity	Electrolysis	Hydrogen / kg	Comparison to Gasoline at 2 gg/kgH$_2$
$0.05	Cost/Effy.	$12.00	$6.00
$0.03	Cost/Effy.	$8.00	$4.00
$0.015	Cost/Effy.	$6.80	$3.40

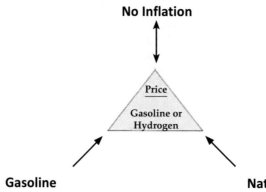

Oil/Barrel	Gasoline/Gallon	Natural Gas/mmbtu	Hydrogen/kg	With CCS
$50	$2	$4	$3.00	$3.50
$120	$5	$8	$3.50	$4.25
$200	$8	$14	$4.50	$5.25

* 1 kg H$_2$ equivalent to 2 gallons of gasoline

* 1 kg H$_2$ in FCV—70 miles 1 gal. Gasoline ICEV—35 miles

With current technology and off-peak power electrolysed hydrogen can be competitive with gasoline and can eliminate greenhouse gases with FCEV's

A realistic cost of off-peak power can be $0.015 / kWh reducing the cost of hydrogen within the range of gasoline. Methane reformed hydrogen includes transportation.

Chapter 6

WORLD'S FIRST PRIVATE COMMERCIAL HYDROGEN PRODUCTION PLANT BY ELECTROLYSIS

Introduction

This hydrogen production plant, located in southern Idaho, was developed in 2006-2007 by independent investors without any collaboration or assistance from utilities or manufacturers. Technical assistance was provided by an engineer from the state university; and a local contractor, under the owner's direction, built the plant. It operates unattended, with software providing all the operational data to an office about forty miles in distance.

Wind turbine generators were installed adjacent to the hydrogen plant to supply the electricity. The plant is also connected to the utility grid.

The hydrogen is marketed to a distributor who takes delivery of the pressurized hydrogen at the plant.

PLANT (See reference below)

Two proton H240 electrolysers, 58% efficiency, 40 kW each and 90% operating capacity, producing 228 scf H_2 per hour each.

Average power usage during ten year life 18 kWh/100 scf of hydrogen. 1kg of H_2 = 422 scf. 4.22 x 18 kWh = 75.96 (76) kWh/kg of H2.

> Compressor 4 kWh
> Auxiliary loads 0.5 kWh.
> Water 1.24 gal. deionized per unit (x 2) per hour.
> Two Entegrity wind systems EW15, total 132 kW connected. (2)
> Capacity factor 24% for wind energy. (2)
> Gross electricity requirement 650,000 kWh/yr.
> Wind energy production 277,000 kWh/yr. (2)
> Purchased electricity 373,000 kWh/yr @ $0.05/kWh. (2)

Note (1): By permission of Tom Griffith, President / CEO, Synthetic Energy Inc., P.O. Box 2247, Ketchum, Idaho, 83340, Ph. 208-727-1991, tomg@syntheticenergy.com http://www.syntheticenergy.com

Note (2): The following financial analyses exclude the wind turbine generators.

Note (3): Synthetic Energy Inc. and Proton Energy Company have a development program to install Proton's new 80 kg/day electrolyzer to recover UHP oxygen for which they have a lucrative market.

PLANT CAPITAL COST

Two PEM Electrolysers, 80 kW @ $130,000 each	$260,000 (1)
On-site costs for electrolysers	100,000 (2)
Site work and building	100,000
Engineering & Project Management	40,000
Total	$500,000 (3)

Notes:

(1) 40 kW of electrolyser capacity @ $130,000 per electrolyzer = $3,250 per kW
(2) Includes automatic and remote operations facilities: no staff on site.
(3) Actual facility included two wind turbine generators, tax benefits, and a grant. Not included in this analysis.

PRODUCTION DATA

Annual Production two electrolysers x 228 SCF/hr. x 7,972 hrs. (capacity factor 91%) = 3,635,050 SCF. Convert to units of 100 SCF = 36,350 CSCF. (1) Convert to kg (2) = 3,635,050/422 = 8,614/kg

Electricity to electrolysers avg. 18 kwh/100 SCF (CSCF) x 36,350 CSCF = 654,300 kWh/yr.

Electricity to compressor and plant auxiliaries 4.5 kWh/h x 7,972 hr = 35,874 kWh/yr.

Deionized water 2.48 gal./hr. x 7,972 hr. = 22,640 gal/yr.

Note:

(1) This commercial plant markets the hydrogen in units of 100 standard cubic feet.
(2) 1 kg of H2 = 38 SCF/90 gm x 1,000 gm/kg = 422 SCF of hydrogen.

PRODUCTION COSTS FOR THIS PLANT (1)

	PerCCSF	Per KG	Annual
Electricity to Electrolysers 654,300 kWh/36,350 CSCF @$0.05/KWh	$0.90		
654,300 kWh/8,614 kg @ $0.05/kWh		$3.80	
654,300 kWh @ $0.05=			$32,715
Electricity for Compressor & plant auxiliaries 35,874 kWh x $0.05	0.05	0.21	1,809
Water 22,640 gallons at $0.265/gallon	0.17	0.70	6,000
General & Administration (incl. compressor o/h and electrolyser stack refurbishment every five years)	0.69	2.92	25,180
Sub-totals operating costs	1.81	7.63	$65,704
Twenty year amortization	0.69	2.90	25,000
Interest at 6% on avg. $250,000	0.41	1.74	15,000
Total costs	$2.91	$12.27	$105,704

Note (1): Hydrogen is sold to a distributor on a profitable basis.

FINANCIAL ANALYSES OF HYDROGEN PRODUCTION COSTS WITH PROJECTIONS OF A RANGE OF FUTURE POTENTIAL INPUTS FOR THE PRIVATELY INVESTED HYDROGEN COMMERCIAL PLANT

INPUTS:

(1) Variable cost of electricity between $0.05/kWh, $0.03/kWh and $0.015/kWh.
- Reduces cost for electricity to the electrolysers from $4.00 per kg to $2.40 and to $1.20.

(2) Cost of electrolysers between $3,250 per kW and $500 per kW.
- Reduces plant capital cost from $500,000 to $260,000.
- Reduces amortization cost from $25,000/yr. to $13,000 and cost per kg from $2.90 to $1.51
- Reduces interest cost from $15,000/yr to $7,800 and cost per kg from $1.74 to $ $0.90

Combining both cost improvements reduces the total cost from $12.06 / kg (with $0.05 / kWh) to $8.43 /kg (with $0.03 / kWh)

(3) Improving the efficiency from 58% to 75% for the electrolyser (an industry target) reduces the consumption of electricity from $4.00 / kg @ $0.05 / kWh to $3.09 / kg and at $0.03 /kWh from $2.40 / kg to $1.86 / kg. Combining this improvement with (1) and (2) reduces the cost to $9.12 / kg (with $0.05 / kWh) and $7.89 (with $0.03 /kWh)

FINANCIAL ANALYSES OF HYDROGEN PRODUCTION COSTS WITH PROJECTIONS OF A RANGE OF FUTURE POTENTIAL INPUTS FOR THE PRIVATELY INVESTED HYDROGEN COMMERCIAL PLANT

The table provides a comparison of cost of production of hydrogen from this commercial plant with three different rates for electricity, for electrolysers at $3,250 per kW vs. $500 per kW and for electrolyser efficiencies of 58% (actual for the plant) and the target 75%.

Inputs per kg of H2				
Electricity $/kWh	0.015	0.03	0.05	0.08
80 kWh / kg	$1.20	$2.40	$4.00	$6.40
Water	0.70	0.70	0.70	0.70
G and A	2.92	2.92	2.92	2.92
Sub-totals	4.82	6.02	7.62	10.02
Electrolysers @ $3,250/kW *	4.64	4.64	4.64	4.64
Total Cost *	9.46	10.54	12.06	14.66
Increased Efficiency 58 to 75%	(0.27)	(0.54)	(0.91)	(1.45)
Total Cost *	9.19	10.00	11.15	13.21
Electrolysers @ $500/kW **	2.41	2.41	2.41	2.41
Total Cost **	7.23	8.43	10.03	12.43
Increased Efficiency 58 to 75%	(0.27)	(0.54)	(0.91)	(1.45)
Total Cost **	6.96	7.89	9.12	10.98

EFFECT OF SENSITIVITIES ON HYDROGEN PRODUCTION COSTS

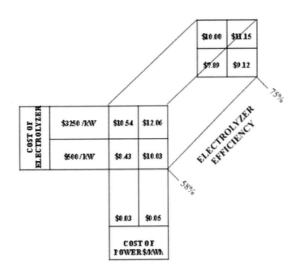

- The effect of reducing the electrolyzer cost from $3,250 to $500 per kW reduced the cost of producing the hydrogen by a little over $2.00 per kg.

- The effect of reducing the cost of electricity input from $0.05 to $0.03 is in the order of $1.50 per kg of hydrogen.

- The effect of improving the electrolyzer efficiency from 58% to 75% is less than $1.00 per kg of hydrogen.

- This suggests that while the combination of all three improvements reduces the cost from hydrogen by over $4.00 per kg the greatest advantage is gained by focusing on reduction of the cost of the electrolyzer.

Chapter 7

HYDROGEN FROM OTHER SOURCES

Hydrogen from Coal

The NHA (National Hydrogen Association) with UND (University of North Dakota) and the Center for Energy and Economic Development has a working group examining the production of hydrogen from coal.

Hydrogen from Nuclear High Temperature Steam

The NHA has a nuclear study group focusing on hydrogen production by nuclear power. The High Temperature Steam Electrolysis (HTSE) process shows promise.

Hydrogen Recovery from Waste Product of Electrolytic Chemical Production Plants

Hydrogen is recovered from chlor-alkali electrolysis plants as a by-product. It requires treatment to remove traces of chlorine and moisture.

Hydrogen from Solar Water Splitting at High Temperature

A European team of scientists have established a pilot reactor at the CIEMAT research center Plataforma Solar de Almeria in Spain, where there is an outstanding level of solar irradiation. In November, 2008, they succeeded in producing hydrogen using solar energy and achieved a yield of 30%. The team has a target of 50%. Their pilot reactor has a solar input rating of 100 kW. A mixed iron oxide, applied to a ceramic honeycomb structure, modifies the course of the reaction and provides for its functioning at 800 degrees Celsius. The team indicates development will require several more years. Ref. Sun & Wind Energy 5/2009.

Hydrogen from Magnesium and Aluminum

This process, under research and development, would permit these two elements to produce hydrogen onboard a vehicle with the probability of an extended range between refueling. It is understood that patents are pending for the process.

Hydrogen from Algae and Waste Streams

Algae reactors have been developed to produce hydrogen with photosynthesis. The process of fermentation of organics has been developed to produce hydrogen. Microbes with some aquatic plants, subjected to an electric current, can produce hydrogen (biocatalysed electrolysis).

Thermomechanical Production of Hydrogen

This method depends on high temperatures in the order of 1000 degrees Celsius, at which water can be split into hydrogen and oxygen. Nuclear high temperature reactors are considered to be a potential and cost-effective source of the heat for economical production of hydrogen.

Thermochemical Production of Hydrogen

There are numerous subprocesses utilizing various combinations of metal catalysts which, combined with heat, will produce hydrogen. These processes are without electrolysis. On a lab scale some are indicated to have a better energy efficiency than high-pressure electrolysis.

Electromagnetic Radiation

This process, being developed by Genesys, LLC, uses a technology referred to as radiant energy transfer. Electromagnetic radiation is used to break the oxygen-hydrogen bond with water in the vapor phase, utilizing only moderate heat.

Chapter 8

CARBON DIOXIDE (EQUIVALENT)

GREENHOUSE GAS (GHG)

EMISSIONS AND CREDITS

Carbon Dioxide Emissions

The National Energy Information Center of the U.S. Department of Energy, in April 2004, projected world carbon dioxide emissions to rise from 23.9 billion (giga) metric tonnes in 2001 to 27.7 billion in 2010 and 37.1 billion in 2025. In this study, 61% of the increase from 2001 to 2025 is expected to be in the developing countries. The report notes that even if the developed world took steps to reduce carbon dioxide emissions, there would still be substantial increases.

The International Energy Agency, in the *World Energy Outlook 2009*, reported CO_2 emissions worldwide at 20.9 gigatonnes in 1990, 28.8 Gt in 2007 and projected to rise to 34.5 GT in 2020 and 40.2 GT in 2030. This is based on governments making no changes to their existing policies.

The IEA 2009 projects a desirable target (some authorities describe it as essential), based on collective action, to limit the long-term concentration of CO_2 equivalent to 450 parts per million (ppm).

China and World GHG Emissions

Leading up to the Copenhagen Conference in early 2010, China indicated the developed countries should reduce greenhouse gas emissions by 40% from their 1990 levels by 2020. The European Union asked for a 30% drop in the same period. A U.S. draft at the time proposed a 5% drop by 2020 from 1990 levels.

China's emissions were 4.6 tons per capita in 2006 compared to 19.8 tons for the U.S. and 12 tons for Russia, according to a U.S. DOE report.

Canada

In Canada, the federal government is moving to a fully market-based carbon price by 2018, when the average price is expected to be about $65 per tonne of CO_2e (Windsight 2008).

European Union Energy Commission

The European Union's Energy Commission has a plan, driven by climate change and energy sufficiency, to obtain 20% of its energy from renewable sources by 2020.

United States Energy Outlook 2010 GHG Emissions

The United States Annual Energy Outlook 2010 (December, 2009) provides a projection of carbon dioxide emissions related to energy consumption. CO_2 from these sources is reported to be 5,814 million metric tons (5.8 giga tonnes) in 2008, 20% of the world total from all sources, increasing an average 0.3% annually to 6,320 million (6.3 giga tonnes) in 2035. The three primary sources of energy consumption are electric power at about 41%, transportation at about 33 %, and buildings and industrial with 26%. The relative mix through 2035 changes less than 1% in each category.

World Energy Technology Outlook for CO_2 Emissions

A World Energy Technology Outlook to 2050 (issued in 2007) reports on two projections of CO_2 emissions. The first is based on a continuation of existing economic and technological trends. It indicates continuing deployment of non-fossil fuels, which temper continuing expansion of coal use. The resulting emission profile shows a concentration of CO_2 in the atmosphere between 900 and 1000 ppm in 2050. This far exceeds what is considered to be acceptable. The second projection is referred to as the carbon constraint case. Renewables and nuclear provide more than 20% of the total world energy demand, renewables providing 30% of electricity generation and nuclear electricity 40%. The objective is long-term stabilization of the CO_2 in the atmosphere at 500 ppm.

An example of the long term concern of many regions of the world with climate warming is found in an exerpt from the

Charlotte Observer. "Water is rising three times faster on the North Carolina coast than it did a century ago as warming oceans expand and land ice melts. It's the beginning of what a North Carolina science panel of experts believe will be a one-metre increase by 2100. About 5,000 square km of its low, flat coast is one metre or less above water."

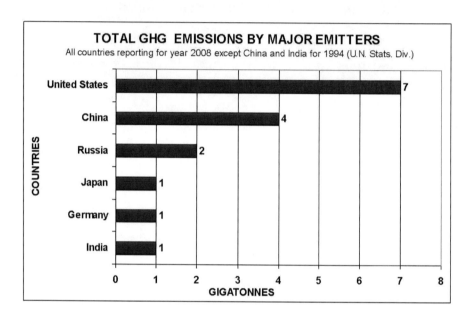

The European Union, by countries, is not provided here. The European Environmental Agency (EEA) has reported total emissions in 1990 of 5.8 gigatonnes and 4.9 gigatonnes in 2008—about 20% of the world total.

BP Alternative Energy and Rio Tinto

This joint venture notes there is an emerging consensus that global greenhouse gas emissions need to be reduced to well below the current levels by the middle of this century.

The power industry is responsible for about 40% of all carbon dioxide emissions, approximately twice that from transport.

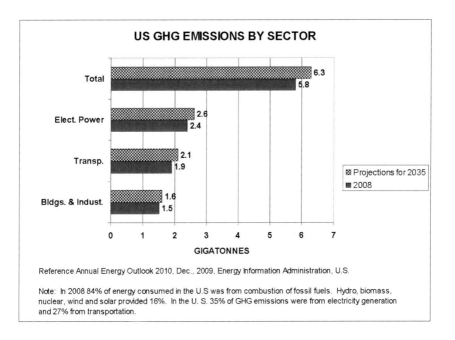

Reference Annual Energy Outlook 2010, Dec., 2009, Energy Information Administration, U.S.

Note: In 2008 84% of energy consumed in the U.S was from combustion of fossil fuels. Hydro, biomass, nuclear, wind and solar provided 16%. In the U. S. 35% of GHG emissions were from electricity generation and 27% from transportation.

Greenhouse Gases in China and Proactivity

In 2008, China became the country with the greatest amount of greenhouse gas emissions, with the United States at number two.

China has set a target of 23% of all electricity to be generated by renewable energy sources by 2020.

In 2009, China became the No. 1 country for most wind energy installations, followed by the United States.

China has the first mass-produced hybrid vehicle which plugs into a home outlet. At the same time, they introduced gasoline mileage standards for automobiles higher than those currently required in the United States.

World Concentrations of Greenhouse Gases

Table of Key Greenhouse Gases Concentration

Gases	Pre Industrial Revolution	Present Levels
CO_2	280ppm	388ppm
CH_4	700ppb	1,745ppb
N_2O	270ppb	314ppb
CFC_{12}	0ppb	533ppd

Notes per the U.S. EPA:

Carbon dioxide is the most important anthropogenic GHG. Over 650,000 years, it has cycled between lows of 180-185 ppm and highs of 285-300 ppm. In the past 150 years, it has risen from about 280 ppm to 387 ppm in 2009.

Methane, CH_4, has varied over time in a range below 800 ppb. Since about 1900, it has increased by 200 ppb the next forty years, then the next twenty years, and then the next 10 years to about 1,400 ppb; and to 1,800 ppb in 2008.

Other GHGs have risen in similar patterns.

Carbon Dioxide from Current Hydrogen Production Method

About 90% of all hydrogen currently being produced is by steam-reforming methane (natural gas). As described in chapter 2, each kilogram of hydrogen produced by this method releases six and one-half kilograms of carbon dioxide to the atmosphere.

World hydrogen usage is currently over forty billion kilograms annually. With 90% by steam-reforming methane, the volume of carbon dioxide released to the atmosphere is

$$0.9 \times 40 \times 10^9 \text{ kg H}_2/\text{yr} \times 6.5 \text{ kg CO}_2/\text{ kg H}_2 = 234 \times 10^9 \text{ kg CO}_2/\text{yr}$$

or 234 million tonnes of carbon dioxide annually.

Forty percent of hydrogen production is used in the refining of gasoline and diesel, 20% for production of fertilizer, and 40% for more than eight other manufacturing processes. Some reports indicate the proportion used for fertilizer production is almost equal to that used for refining.

A combination of increasing market price for natural gas and significant pricing of carbon dioxide as an offset or tax will bring the cost of hydrogen by steam-reforming methane up to or exceeding the cost of hydrogen produced from renewable energy with electrolysis. Consistent policies and incentives going forward for development of renewable energies will bring mass production with competitive costs. The alternative of hydrogen from wind energy and electrolysis would virtually eliminate this source of carbon dioxide.

Purchase of Carbon Dioxide Offsets

Markets are developing for the purchase of carbon dioxide credits to offset GHGs being unavoidably produced in certain industries and businesses. The cement and lime industries are examples where production of carbon dioxide is unavoidable. They use calcium carbonate, $CACO_3$, as their basic material source and use heat in a kiln to separate the CO_2. It is reported that one tonne of carbon dioxide is emitted for every tonne of cement produced. Yet cement and lime appear to continue to be very essential building components.

Alberta has an active GHG offset program. Canwea June 2010 reports that Alberta companies purchased 3.8 million tonnes of greenhouse offset credits in 2009 to help comply with the province's legislated emissions reduction goals. About 11% of the total, or 407,000 tonnes came from wind energy projects. Price is in the range of $12 to $14 per tonne of CO_2e.

Electric Power Generation GHG Emissions

There are clear options in the electricity generation industry to reduce GHGs. Coal (lignite) power plants emit about 1,000 kg CO_2e/MWh. Combined cycle gas turbines (CCGTs) emit about 350 kg CO_2e/MWh, or one-third that of coal-fired plants.

The combination of CCGTs operating with compressed air energy storage systems supplied from renewable energy sources will reduce the GHG emissions to about 125 kg CO_2e/MWh. This would be about one-sixth that of coal-fired plants.

United Nations Food and Agriculture Organization Claims of GHG Emissions from Beef Cattle

The UN FAO has reported beef cattle, by emission of methane, are one of the greatest contributors to GHGs. *Fortune* magazine of April 12, 2010, noted that the FAO reports meat accounts for 18% of the world's greenhouse gases every year, compared to 13% for vehicles, with beef singled out as the greatest "culprit."

However, it can be shown that despite the emission of methane, the overall life cycle of beef cattle is marketedly beneficial to the environment with significant reduction of GHGs in the atmosphere.

Consider the life-cycle Greenhouse Gases (e) (GHG's equivalent) of beef production. The dry weight feed to cattle is about forty pounds per day, depending on the size of the cow and average ambient temperature. Allowing for moisture at 10% and ash at 5% (both possibly high) of this forty pounds of daily feed, about thirty-four pounds is volatile matter based on carbon. When dry grass burns, it leaves only a filmy black dust of ash. Photosynthesis extracted this carbon from the carbon dioxide (read GHG) in the atmosphere to grow the grass with mineral and other nutrients from the soil. The grass or hay converted carbon dioxide in the atmosphere into carbon in the grass and oxygen in the air.

40 lbs. x (100%—(10% moisture + 5% ash) = 40 lbs. x 85%
= 34 lbs.

Carbon dioxide has a molecular weight made up of carbon at twelve and oxygen at sixteen. As there are two oxygen atoms in carbon dioxide, the portion of oxygen is 32 compared to 12 for the carbon, and the ratio of carbon to carbon dioxide is twelve to forty-four.

$$CO_2 = C + 2O = 12 + 2 \times 16 = 44$$

The thirty-four pounds of volatile matter per day consumed by the cow is from grass or hay. The volatile matter consists of about 50% carbon or seventeen pounds per day (see table below). This carbon, by photosynthesis, took forty-four divided by twelve multiplied by seventeen pounds equals sixty-two pounds of carbon dioxide from the atmosphere. Every day the cow consumes its forty pounds of grass or hay, the field regrows its replacements and takes sixty-two pounds of carbon dioxide from the atmosphere while releasing the oxygen back to the atmosphere.

34 lb x 50% C (carbon portion) x 44 CO_2 / 12 C = 62 lb CO_2

Now consider the methane emitted to the atmosphere by the cow, a major criticism of the cattle industry. It is known that the greenhouse gas effect on climate warming is much greater from methane than from carbon dioxide. The multiplier is twenty-one times. So one pound of methane has the same effect as twenty-one pounds of carbon dioxide. How many pounds of methane does a cow emit per day, and how does this compare to the amount of carbon dioxide its replacement feed takes from the atmosphere? It may be noted that methane breaks down in the atmosphere in eight years; however, carbon dioxide does not.

The Journal of Animal Science, 1995 (73-2483-2402) reports a range of emissions from cows of 250 to 500 liters of methane per day. The density of gaseous methane is 0.7167 grams per liter. Taking an average of the preceding reported levels of emission at 375 liters per day the weight would be 375 liters x 0.7167 grams/liter = 269 grams per cow per day. Converting this to pounds with 2.2 pounds per 1,000 grams the methane emissions per cow per day would be 269 / 1000 x 2.2 = 0.591, say 0.6 pounds. With methane more critical as a greenhouse gas than carbon dioxide by twenty-one times this will be the equivalent of 21 x 0.6 = 12.6 pounds of carbon dioxide equivalent per day per cow. Compare this with the sixty-two pounds of carbon dioxide indirectly removed from the atmosphere by a cow eating its grass and hay.

375 liters / cow/day x 0.7167 grams/liter = 269 grams/cow/day
269 grams/cow/day / 2.2 lbs./1,000 grams = 0.591, approx. 0.6 lbs./day/cow
0.6 lbs/cow/day NH_4 (methane) x 21 CO_2(e) = 12.6 lbs./cow/day CO_2(e)

Chapter 9

DISTRIBUTED POWER BY HYDROGEN FUEL CELLS

AND

COMPRESSED AIR ENERGY SYSTEMS

From little acorns,

    mighty oaks will grow.

the consumption of fossil fuels and production of GHGs. As fossil fuel supply is finite, especially for oil and gas, the accelerating demand will exceed supply and create stresses globally. This rapidly looming catastrophe can be avoided by the peaceful accelerated development of renewable energies which do not create regional tensions and are infinite.

Manure's carbon-to-nitrogen ratio is a key factor in making nitrogen available to plants, because it drives microbial decomposition (Alina Rice, Washington State).

A carbon credit per cow resulting from its contribution to reduced greenhouse gases with organic agriculture *can be added* to the credit from its consumption of grass and hay.

Table

Sources for analysis of grass or dry hay are not readily available. A comparison may be made with an analysis of dry bagasse (Mark's Mechanical Engineers Handbook):

Carbon	44% (fixed carbon 12%)
Hydrogen	6%) Volatile matter
Oxygen	47%) and fixed carbon 97%
Ash	2.5%
Nitrogen	0.5%
Ash	2.5%

Observations and Conclusions

- The preceding analysis is one example of the serious conflicting scientific reports concerning greenhouse gas emissions and the cumulative effect on the climate.
- There is clear evidence that the various components making up GHGs are accumulating in the atmosphere.
- There are reports by scientists which correlate the history over hundreds of thousands of years of accumulation of GHGs coincident with warming of the earth.
- There appear to be other sound scientific reports which refute the theory of GHG accumulations resulting in climate warming.
- Fossil fuels are shown to be the major contributor of GHGs to the atmosphere. The combination of global improvement of standard of living and population growth are resulting in increased generation of electric power which is accelerating

Greenhouse gas credits of nearly fifty pounds of carbon dioxide per cow per day; 18,000 pounds per year; eight metric tonnes per year are attribual to each beef animal.

$$62 \text{ lbs } CO_2/\text{day} - 12.6 \text{ lbs } CO_2 \text{ (e)}/\text{day} \times 365 \text{ day}/\text{yr} / 2,200 \text{ lbs}/\text{tonne} = 8 \text{ tonnes}/\text{yr } CO_2.$$

Caked cow "dung" is used as a fuel to cook food and provide heat in parts of Asia and Africa. It is an excellent fuel due to its high volatile carbon content—the carbon which came from the carbon dioxide in the atmosphere by photosynthesis. In some remote parts of Asia Minor, a family needs at least four cows to be able to collect enough of this fuel to last through their winter. While not of great magnitude in terms of greenhouse gases, this cycle is close to neutral because this fuel originates with pasture grass taking carbon dioxide from the atmosphere, and this greenhouse gas is returned to the atmosphere when it is burned as fuel.

An FAO Corporate Document on organic agriculture, environment, and food security notes that "carbon dioxide emissions per hectare of organic agriculture systems are forty-eight to sixty-six percent lower than in conventional systems. Carbon dioxide emissions of German organic farms were calculated to be 0.5 tonnes per hectare whereas in conventional agriculture the amount was 1.3 tonnes, a difference of sixty percent (Haas and Kopke-1994)." The main effects of organic agriculture that are responsible for this difference are:

- The maintenance and increase of soil fertility by the use of farmyard manure
- The omission of synthetic fertilizers and synthetic pesticides
- The lower use of high energy consuming feedstuff.

Organic agriculture enables ecosystems to better adjust to the effects of climate change and offers a major potential to reduce the emissions of agricultural greenhouse gases. It has a high potential to counter soil degradation as it is more resilient both to water stress and to nutrient loss.

Hydrogen Energy Storage to Convert Variable Renewable Electricity into Peak Power at Load Centers

Distributed Power by Fuel Cells

Existing transmission and distribution facilities are overloaded at times of peak demand.

Building new transmission and distribution is becoming extremely difficult to permit (NIMBY—not in my backyard) as well as being costly for the **incremental** additional power to be moved.

Combined electrolysis and fuel cell unit plants can be located at load centers in the form of Distributed Power Centres (DPCs).

Alternatively, methane supplied fuel cells can supply peak power demand during daytime hours when natural gas heating requirements are lower. Efficiency when utilizing the heat from the methane to hydrogen conversion and the fuel cell hydrogen to electricity conversion is exceptionally favorable, over 70%, with some heat recovery.

During low-demand nighttime periods, the existing generation, transmission, and distribution (GTD) facilities can carry exceptionally low cost power to the DPC where it is converted to hydrogen and stored.

When daytime peak-demand periods return, the hydrogen is supplied to the fuel cells to return electric power to the grid.

It is desirable for the utility, in a regulated system, to do a real incremental costing for the low-demand period power and also for the high-demand period power; the latter with DPCs, including the advantage of being distributed power which also avoids the investment in additional GTDs to provide only a small amount of peak power.

Currently available electrolyser-fuel cell systems have a global efficiency of 25%. Research and development is indicating a potential global efficiency as high as 33% to 45% will be achievable.

Low demand power at 1 to 2 cents or less per kilowatt-hour can economically feed back high-demand distributed power at 6 to 12 cents per kilowatt-hour and higher at the load centers with present DPC efficiencies, including the capital investment.

Using renewable electricity to supply the distributed power unit at low-demand periods will ensure negligible greenhouse gas emissions and sustainability.

The Global Economics of DPCs on Utility Systems Financial Investment Strategy

The global economics of DPC's, in addition to conversion of very low value off-peak electricity to hydrogen and returning it to the grid through fuel cells (or internal combustion engine-generators (ICE-Gs) during high-value, peak-demand periods must include consideration of the potential for DPCs to reduce the daytime peaks sufficiently as to defer or eliminate the need

- to increase distribution system capacity,
- to increase and build new transmission capacity,
- to add new generation facilities,
- for additional spinning reserves otherwise considered essential,
- for standby and support electricity in case of disruption of the main electrical supply system (blackouts or curtailments).

The global economics of the impact of a major installation of DPCs on the total electrical supply system can be evaluated and will serve to further justify the investment.

Peak Demand Curves

Load curves providing the measure of electrical demand vary over a twenty-four-hour day, and the pattern will vary with the time of year and the characteristics of each load center. As an example, a downtown core of office buildings and commercial shops will have a very low demand during nighttime hours and a fairly heavy steady demand during the daytime, except on weekends. Residential areas generally have a very typical demand curve but varying with the seasons. Locations, capacities, and effectiveness of distributed power centers will be determined by the demand curves for each particular location, with optimization for seasonal and other characteristics affecting the demand for that location.

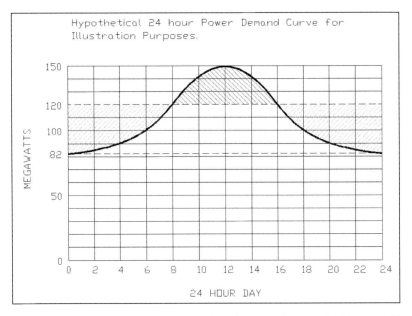

The horizontal line drawn across the chart at the peak demand level of 150 MW shows the 24 hour availability which is required of the Generation, Transmission and Distribution (GTD) supply facilities.

The second horizontal line drawn across the chart at 120 MW illustrates how a section of peak demand may be replaced by DPC's which could be supplied by electricity from the cross-hatched areas below this line, then returned to the grid to displace the peak demand sector above the line.

The quantity of electricity above the 120 MW level is approximately 30 MW x 8 hrs. x 50% = 120 MWh. The quantity of electricity available between the 120 MW and the 82 MW levels, but not utilized, is approximately 38 MW x 16 hrs. x 50% = approximately 300 MWhrs. In this hypothetical example the unused GTD facilities will supply DPCs, and the DPCs then supply the peak electricity requirement between 120 and 150 MW. In this case, the GTD facilities for this peak 30 MW are not required, a major capital investment and operating cost avoided. The illustration used here assumes a targeted 40% global efficiency is achievable.

The capital investment in the 38 MW of DPCs (at the targeted efficiency of 40% and *targeted costs*) will be

    Investment in the electrolysers
    38 MW x $500,000 / MW   = $19,000,000
    Investment in fuel cells
    30 MW x $500,000 / MW   = $15,000,000

    Total $34,000,000

The electrolysers and fuel cells are currently produced manually. Increased demand is necessary to move the process to mass production which will reduce the combined costs below $1 million per MW.

The capital investment in the GTDs, which is avoided for the 30 MW of eliminated peak demand, is in the order of $5 million per MW or $150 million.

The DPCs supplying electricity to the grid during daytime peaks can also supply building heating or cooling, which improves the overall efficiency.

In this example, the electricity drawn from the grid during the off-peak period should be at a low value, in the order of 1.5 cents per kWh. It can then be returned to the grid during the peak period at 1.5 cents x 0.4% efficiency = 3.75 cents per kWh plus recovery of the investment. Its real value would be more in the order of 10 to 20 cents per kWh or higher. A gain of 6 cents per kWh on 120 MWh/day would be $60.00 per MWh x 120 MWh/day x 365 days/yr. = (approx.) $2.60 million annually.

Operating costs for the DPC system would be considerably less than for the GTD facilities it would replace. The DPCs operate without attendants and are managed remotely from the utility's central control station.

If natural gas is used in the DPC to produce the hydrogen by reforming and if the heat from the combined process is used to meet building requirements, then the overall efficiency is in the order of 70%. No other system can claim such a high efficiency. By using this system to replace electricity from any form of thermal process will effect a significant reduction, but not eliminate, greenhouse gases.

If the off-peak electricity to the DPCs is from renewable electricity, then there is virtually no greenhouse gas emission when supplying the peak demand. The value of variable renewable energy during peak periods and its value to low-demand periods should be established. Supplying the low-demand period renewable energy to DPCs is a major step in dealing with its variable nature.

Interconnection between utility grids can assist in supplying the peak demand sector without having to provide sufficient generating capacity within its own system. An example is a hydro system, especially rich in capacity but not in energy, supplying to a neighboring grid, which is heavy to thermal generation, during daytime peaks, and then drawing energy from the thermal facilities during low-demand periods. This accommodation does not, however, reduce the peak-demand requirement of the transmission and distribution facilities.

DPCs With Natural Gas

DPCs utilizing natural gas with a fuel cell, with heat recovery, will have an energy efficiency of over 70 %; 47% without heat recovery. A carbon tax or cap and trade cost for the considerable carbon dioxide emissions from the steam reforming of the natural gas to produce the hydrogen for the fuel cell would be a cost disadvantage of these DPC's.

Development of Distributed Power Centers

- Arizona Public Service (APS) target 1,500 MW from DPCs located near customers with sources including rooftop solar, wind, biogas, biomass, geothermal, hybrid wind, renewable natural gas, and hydropower technologies. The APS's renewable target by 2025 is 40%.
- Pepperidge Farm in Connecticut has installed at 1.2 MW fuel cell at a cost of $6 million of which $3.5 million was a grant. The full cost of plant is $5,000 per kW. The plant generates electricity by converting natural gas to recover hydrogen which is converted in the fuel cell to electricity. Heat from the conversion is used in the bakery. Water is also a pure byproduct.
- FuelCell Energy Inc., Danbury, Connecticut, manufactures direct fuel cells (DFCs) with capacities of 300 kW up to 2.4 MW, and these can be expanded to 50 MW. They use natural gas, propane, and biogas. The net electrical efficiency is 47%, not including any heat recovery, better than almost all types of fossil-fueled thermal power plants and provide the advantage of power generation at the load center.
- Enbridge Inc. and FuelCell Energy Inc. have opened what they are calling the world's first power plant to pair fuel cells with an emissions-free process that converts waste energy into electricity. Opened in October 2008, the 2.2 MW plant cost $10 mm. It produces electricity through a chemical reaction, with a turbo-expander which harvests the primary energy from natural gas to generate power using a turbine. It is reported that the clean benefits of FuelCell Energy's fuel cell are linked with recovery of wasted pressure that also produces useful electricity. (National Post, October 24, 2008).
- Posco Power, Korea is developing 38 MW of fuel cell distributed power systems.
- Sierra Nevada Brewing Company in Chico, California, operates a 1 MW solid fuel cell plant fueled by gasses from the beer process with some natural gas.

- Ford Motor Company in Ontario, Canada, is installing fuel cells for 300 kW, using fumes collected in their auto painting plant as the energy source.
- CSU (California State University) Northridge in California operates a 1 MW fuel cell electric power and heating plant, utilizing four 250 kW molten carbonate fuel cells. The project is considered to be financially viable and reduces greenhouse gas emissions by about 70% compared to existing California power systems. By using the waste heat from the hydrogen generation fuel cell operation, the electrical efficiency of 47% is improved to an overall efficiency of about 80%.

Honda has developed a Home Energy Station, with refinements, since 2003. It generates the hydrogen from natural gas and through fuel cells provides co-generation. It provides heat and electricity for the home and hydrogen for fuel-cell vehicles. The cogeneration facility offers almost 50% efficiency, the same as that achievable with a combined cycle gas turbine (CCGT) plant.

- British Columbia Hydro, Canada, is developing its HARP project (hydrogen assisted renewable power) in an isolated community of Bella Coola. An existing run-of-river hydro plant is capable of producing an excess of electricity during nighttime low-demand periods, and if this is not utilized, the water flow is wasted. This nighttime power generating capability is to be used by electrolysis to produce and store hydrogen. During daytime peaks, when the hydro plant cannot meet full demand, the hydrogen—through fuel cells—will supply part of this peak.

DPCs with Variable Renewable Energy Sources

Integrating variable renewable electricity into the grid requires "smart-grid" management. Most utilities depend on thermal and nuclear power generation which operates best when its facilities are base loaded. However, these plants must adjust their output to accommodate the variable renewable energy when its facilities are producing. Incorporating DPC systems enhances the value of variable renewable energy and reduces the operating rate adjustments otherwise required of the thermal and nuclear plants.

Capacity Utilization Factor (CUF) of Thermal—and/or Nuclear-Based Utilities

Wind, solar, and run-of-river variable renewable energy sources evaluate their output with a capacity factor (CF). Wind energy, as an example, varies with the site and the season and, to a lesser extent, with the efficiencies of the wind turbine generators (WTGs). Some wind farms have a CF as low as 20% and others as high as 40% (offshore even higher). CF is the quantity of electricity generated by a wind farm or WTG, in a year, divided by the product of the nameplate capacity of the WTGs and the number of hours in the year. Note that the energy input cost is zero and there is no inflation.

Thermal and nuclear power plants have a capacity factor better than 85%, allowing for periodic maintenance and unscheduled downtime. However, of greater significance is their capacity utilization factor (CUF). Where some part of these facilities are required to meet short-term peak demands, such facilities will have a CUF of 30% or even less. If spinning reserves are included, the CUF for them would be less than 20%. These low CUFs for thermal and nuclear power plants result in a portion of their output being very costly. Standby generating capacity, which is rarely used, reduces the overall CUF and increases the cost for assured reliability. The low CUFs for thermal and nuclear power plants for a portion of their output is very costly. Evaluation of the alternative of DPCs is desirable.

Distributed Power Centres (DPCs) to Reduce New Transmission Investments

As wind energy continues to be developed at an ever-increasing rate, the apparent need for new transmission facilities investment to move it from generally more remote regions to load demand centers is placing emphasis on these funding requirements independent of the wind farm investments.

An interesting example of the potential for a combination of renewable energy storage and distributed power centers to defer or eliminate the need for a major new transmission facility is found in the province of Alberta. Alberta has a market-based electricity system without the rates being designated by some form of utility regulator. Alberta was the first province in Canada to connect over 500 MW of wind energy, all in the southern region of the province. Concern with integrating more wind energy caused the system operator to place a cap on it, since removed. There is reported by CanWea the potential of 12,000 MW of wind energy available and readied for development in Southern Alberta. The province's total power generation is about 12,000 MW, mainly coal-fired plants and natural gas turbines, with coal predominating.

Generation is located primarily in the central part of the province with some in the northeast oil-sands region. There has been rapid population and industrial growth in Alberta with peak electricity demand rising over twenty years from 5,000 MW to 12,000 MW. With most of the generation in central and more northern sectors, there has been a need recognized to provide a major addition to the existing heavily loaded transmission lines supplying power to the southern part of the province. There appear to be some difficulties in clearing obstacles to its construction, which is a common problem, usually NIMBY, shared almost everywhere.

A combination of distributed power centers and wind energy development with energy storage in the southern part of the province could eliminate the need for this new transmission line.

The Alberta Electric Service Operator (AESO) has approved and underway over one billion dollars for transmission system improvements. Additional planning includes north-to-south transmission system reinforcements of over three billion dollars. The Southern Alberta Transmission Reinforcement, at two and one-half billion dollars, was approved in September 2009 to integrate 2,700 MW of new wind energy in Southern Alberta.

Offset by DPCs—A Simple Conceptual Illustration

It assumes the north-south reinforcement will provide for an additional 3,000 MW of electricity to Southern Alberta during peak hours assumed to be 8 hours per day with an average supply of 50% of the peak or 1,500 MW x 8 hours = 12,000 MWh per day. For this illustration, electricity is supplied from the north-central generating facilities during *off-peak hours* when the cost is basically only the variable costs of fuel and the existing north-south transmission lines have capacity available.

The example applies the worst case global system efficiency for DPCs of only 25%, while planners and developers are targeting 40% or better.

In terms of operating costs, DPCs can function without operating staff, controlled remotely by software. Maintenance is low. By comparison, the cost of operating and maintaining the otherwise-required generation, transmission, and distribution (GTD) facilities are considerably higher.

The resource cost for electricity into the DPC's relative to the electricity back to the grid will have an input cost for the off-peak power of about $15 per MWh and a back-to-grid value in peak hours of about $100 per MWh.

> Resource input cost (based on 25% effy.) 4 x 12,000 MWh = 48,000
> MWh/day x $15/MWh = $720,000 per day
> Grid value recovery 12,000 MWh/day x $100/MWh = $1,200,000 per day.
> Annual Gain on resource cost 365 days x $480,000/day = $175,200,000/yr.

Investments avoided by the DPC systems for the North-South Transmission Reinforcement is indicated to be $3 billion. New conventional generation avoided, no longer needed to supply the peak portion of demand, of 3,000 MW at $2 million per MW would be $6 billion. If the new generation would be coal fired (lowest operating cost) with carbon sequestration, the avoided cost would be more than $3 million per MW or $9 billion. Avoided generation and transmission investments could be in the order of $12 billion or greater. If new generation is with combined cycle gas turbine generators (CCGT's), the avoided investment cost for 3,000 MW would be in the order of $3 billion added to the avoided cost of the transmission lines of $3 billion for a total of $6 billion.

Investments for the alternative of DPCs will depend on the growth in demand for electrolysers and fuel cells to the level involving mass production, which will dramatically reduce their costs. The costs here assume mass production will reduce them to close to the targets set by the U.S. Department of Energy, below $350 per kW for electrolysers.

Electrolysers required to convert the off-peak electricity to hydrogen operating over sixteen hours per day will have a capacity of

$$12{,}000 \text{ MWh} \times 4 \text{ (25\% effy)} / 16 \text{ hours} = 3{,}000 \text{ MW}$$

At a unit cost of $400,000 per MW the investment would be $1.2 billion The fuel cells required to convert the hydrogen to electricity fed back to the grid during the peak-demand period will be, based on the targeted unit cost of $500,000 per MW

$$3{,}000 \text{ MW} \times \$500{,}000 \text{ per MW} = \$1.5 \text{ billion}$$

The total investment for the DPCs would be in the order of $3 billion, utilizing off-peak electricity from existing generating plants in the north-central region.

This DPC plan could incorporate vastly increased wind energy development in Southern Alberta with part or all of the investment cost offset by eliminating the need for additional north-south transmission. A portion of the wind energy could be stored utilizing the compressed air energy storage system (CAES) combined with CCGTs for a stored energy efficiency of

85%. Southern Alberta is well sited with considerable cavern storage capability. Utilizing more of the available wind energy in Southern Alberta with storage and CCGTs would greatly reduce the greenhouse gas emissions from existing coal-fired generating plants.

Transmission System Costs Relative to Offsetting DPCs

In Texas, a transmission system is planned at a cost of $5 billion to serve 11,500 MW of wind energy. At about $2 million per MW of installed wind energy capacity, the generating facilities would have a cost of $23 billion. The ratio of cost of the transmission facility to the cost of the generating facilities is indicated to be in the order of 22%. Transmission systems for the main purpose of carrying wind energy have a utilization factor of 25 to 40% depending on the wind resource C.F. Use of some wind to hydrogen energy storage may be justified to reduce the cost of transmission.

Southern California Edison is planning a transmission facility at a cost of $5.5 billion to serve a planned 7,000 MW of wind energy, which would cost in the order of $14 billion. The ratio would be 40%.

The Alberta Electricity System Operator (AESO) is planning a $1.8 billion transmission loop to add up to 2,700 MW of wind power capacity before 2020. Wind energy investment for 2,700 MW would be in the order of $5.4 billion. This indicates a ratio of transmission requirements to wind energy investment of 30%. Other regions indicate a range around 25%.

Observations and Conclusions

- Where suitable cavern storage or other subterranean storage is available, the efficiency of CAES energy storage is most attractive combined with use of the compressed air in combined cycle gas turbine (CCGT) generating plants.
- When CAES is not an available option, the use of DPCs utilizing off-peak power or natural gas can provide a practical system to avoid new GTD investments.

- CAES or EFC DPCs offer sound systems to complement wind energy quantities, which exceed normal grid control facilities.
- Distributed power centers and CAES—when available at load centers—improve electrical system reliability, are a favorable means of firming power supply from variables, maximize the use of existing transmission and distribution facilities, and avoid the need for new GTDs.

Chapter 10

THE POTENTIAL OF RENEWABLE ENERGIES

THREE WORLD CRISES

CLIMATE WARMING

LIMITED CRUDE OIL SUPPLY

LESS AVAILABLE WATER

Winston (Win) Stothert

RENEWABLE ENERGY SOURCES

WILL

REDUCE GREENHOUSE GAS EMISSIONS,

ELIMINATE THE NEED FOR OFFSHORE OIL (FOR THE UNITED STATES AND OTHERS),

REDUCE NEED OF WATER FOR THERMAL PLANTS AND REFINERIES.

Proven technology exists to develop renewable energies on a socially responsible basis:

- to provide hydrogen to replace overseas oil which is supplying gasoline for vehicles

- to provide the electricity which is now supplied by coal fired power plants

- to reduce the demand for water which is now becoming in critical demand

Governments hold the key to changing the mix of energy investments. Policy and regulatory frameworks established at national and international levels will determine whether investment and consumption decisions are steered toward low-carbon options. Ref. IEA-2009

Sustainability of the Energy Systems

The European Union is committed to development of a hydrogen-including economy using hydrogen as an energy carrier.

Hydrogen, unlike electricity, can be stored in small or large quantities for long periods with negligible losses.

It can be produced from many resources including renewables, nuclear, and fossil fuels, and as a by-product of electrolytic chemical plants.

Variable renewable energy can be converted by electrolysis of water to hydrogen, a form of energy storage.

It can be transported by pipelines, and it can be stored in caverns en route just as natural gas has been intermittently stored for many decades. Hydrogen pipelines extending for many hundreds of miles, with underground cavern storage similar to that used for natural gas, have been operating successfully for decades. It can be transported in existing natural gas pipelines as a minimum hydrogen to methane mix referred to as hythane. Specifications for hydrogen pipelines include provisions to counteract embrittlement and leakage.

Alternative systems for storing electricity have been developed and have been in use for many decades. These include hydro storage which is limited regionally, pumped hydro storage dependent on special geographic features, batteries and flywheels which are capacity limited, molten salt underground, compressed air in subterranean storage especially suited to gas turbine generators which provides almost unlimited capacity and others with varying degrees of applicability. All have some efficiency loss.

Hydrogen's Versatility

Hydrogen is used to directly fuel both internal combustion engines (ICEs) and fuel-cell electric vehicles (FCEVs). On-board storage has been improved to provide acceptable travel distances between refueling. The energy content of one kilogram of hydrogen is the same as in one gallon of gasoline; however one kilogram of hydrogen will drive a fuel-cell vehicle twice the distance that can be achieved with one gallon of gasoline in an ICE due to the lower energy conversion efficiency of an internal combustion engine. For this reason, aside from the excessive greenhouse gases of a gasoline driven vehicle, the economic value of a kilogram of hydrogen can be considered double that of a gallon of gasoline in vehicles. Note: The efficiencies of ICEs is experiencing considerable improvement to meet state and federal targets.

Hydrogen's largest use today is to refine oil to gasoline or diesel.

Hydrogen's second largest use today is to produce the most common fertilizer, ammonia (NH_3). In the early 1900,s some of the hydroelectric plants, such as Norsk in Norway and Shawinigan Hydro (Chemical) in Quebec, had huge excesses of virtually free electricity which they used to produce hydrogen and then ammonia fertilizer via an electrolysis process. Norsk's long experience in electrolysis explains why they produce an electrolyser of considerably higher capacity than available from other manufacturers; 485 Nm^3/hr compared to the next highest reported at 60 Nm^3/hr. (Ref. NREL (DOE) Summary of Electrolytic Hydrogen Production, Sept. 2004).

Currently hydrogen is mainly obtained by steam reforming methane (natural gas), but this results in over six kilograms of carbon dioxide being released to the atmosphere for every kilogram of hydrogen produced.

Hydrogen can be produced economically by electrolysis at a load center when the electricity cost is low during low-demand periods with excess capacity available from the generation, transmission and distribution facilities. This stored energy can then be returned to the grid at the load center through a hydrogen fuel cell. The electrolyser-fuel cell facility is noise and vibration free and does not require an attendant. The global efficiency of such a facility is improving rapidly while the investment cost is being dramatically reduced with increasing manufacturing levels. Note: The electric utilities are encouraged to better determine the values including avoided investment costs of power at high-demand periods compared to low-demand periods with the extra value that electrolyser-fuel cell systems located at load centers can provide.

There are numerous other critical and essential uses for hydrogen, the world's most versatile element. Space travel and satellites have only been possible by the use of the high energy of hydrogen as the rocket fuel.

Potential Renewable Energy Worldwide

Researchers at Stanford University did an evaluation of the global potential of wind power. Using only 20% of their estimated economical wind power generation they concluded that wind energy could satisfy the world's electricity demand seven times more than the quantity being consumed in the year 2000. The German Advisory Council on Global Change in 2003 calculated the global technical potential for wind energy from onshore and offshore was 278,000 TWh (terra-watt hours) per year. They then assumed less than 15% would be realizable in a sustainable fashion at 39,000 TWh per year, which was more than double the then current global electricity demand. (Reference GWEC wind energy outlook 2008).

Potential Renewable Energy on a Socially Responsible Basis in the United States (Ref. "NREL-DOE"—Feb. 2007).

The United States provides a good example of what can be accomplished in a high energy production-consumption society with 4% of the world's population and consuming 25% of the world's energy. This can be extrapolated globally.

The U. S. National Renewable Energy Lab, Department of Energy (NREL—DOE) in Colorado has reported their evaluation of the energy potential from the main renewable energy sources in the United States, measured in terms of kilograms of hydrogen. This adjusts for efficiencies and intermittence while reducing the quantities to a common measure. In practice, some of the electricity from renewables will be supplied directly to the grid. This study does not suggest that the only route for storage of variable renewables is hydrogen. Long-term and large-capacity storage at reasonable cost and good efficiencies indicate compressed air energy systems (CAES) are a favorable alternative where suitable conditions exist.

Electricity to Hydrogen Conversion Efficiency and Electrolyser Capacities

The NREL uses an average conversion efficiency of market available products for electrolysis of 58.8 kWh of electricity per kg of hydrogen. This is for the entire system which includes the AC/DC converter and, electrolyser cooling. Norsk Hydro (StatOil) report a similar global system efficiency except theirs includes a water injection screw compressor followed by a reciprocating compressor to bring the hydrogen to 480 psi. Norsk and one other manufacturer report a global system conversion efficiency of 53.5 kWh/kg of hydrogen. This is 9% higher efficiency than the average reported by NREL. Note: In terms of capacity per the NREL Feb. 2007 report Norsk produce electrolysers at 485 Nm3/hr (43.6 kg H_2/hr) which is an eight times larger capacity than the next largest reported by NREL.

The numbers of electrolysers required as a significant part of major electricity systems has been questioned. A comparison is made here of the number of electrolysers which would be required to provide for the output of a typical wind farm with one hundred WTG's of one and one-half megawatt nameplate capacity. Note: Total averaging of the wind farm capacity would not generally be required as some component of the wind farm output would always be acceptable to the grid. The Norsk high capacity unit is considered in this illustration.

1 WTG of 1.5 MW capacity = 1,500 kW (@ 30% C.F. = 450 kW and per hour 450 kWh

however the electrolysers must be able to handle full 100% output.

1 Norsk 43.6 kgH$_2$/hr unit x 53.5 kWh(e) = 2,333 kWh(e)
100 WTG's @ 1.5 MW = 150 MW = 150,000 kW
No. of Norsk Electrolysers required = 150,000 kW/2,333 kW(e) = 64 units.

Allowance for some of the wind farm output always accepted by the grid directly would reduce the number of electrolysers by something in the order of 30% depending on the wind resource and load demand characteristics.

No. of Norsk electrolysers per 100 MW of connected wind farm would be 0.70 x 64 units = 45 units

Wind Resource in the United States.

The NREL report based its evaluation on technologically economic wind resources with class 3 average annual wind speeds and above, at 50 metres above sites with slopes not over 20%. (See note). Production capability was based on 5 MW per km^2, with capacity factors ranging from 0.25 to 0.30 (a very good C.F. is 40%). (See Note). They avoided parks and other socially sensitive areas including distance to habitations. Offshore wind potential was excluded. Wind is the dominant renewable energy resource in the central states.

Using NREL's average electrolyser system efficiency value of 58.8 kWh per kg of hydrogen (a 63% electrolyser system efficiency—hhv) the potential annual production is 273 million tonnes of hydrogen.

Applying the Norsk more favorable electrolyser system efficiency of 53.5 kWh per kg the potential for hydrogen from wind energy would be 300 million tonnes annually.

Note: Class 3 wind at 50 metres height from ground is defined as having an average annual wind speed of 6.4 to 7.0 metres per second or 14.3 to 15.7 miles per hour. Some wind turbines are now being installed with a hub height of 100 metres. The wind becomes stronger at higher elevations.

Note: Capacity Factor (C.F.) is defined as the actual electrical output of a wind turbine generator (WTG) over a year divided by the amount of electricity which would have been generated if the output was at the nameplate capacity of the WTG.

Solar Resource in the United States.

The NREL study based the solar resource on nontracking flat-plate collectors oriented toward the south at latitude tilt. Estimates of average daily total global radiation on these collectors were based on meteorological data. Values ranged from 2.2 kWh/m^2/day in Alaska to 7.0 kWh/m^2/day in parts of the southwest United States.

There were similar exclusions for these as for the wind sites. The study assumed only 10% of specified areas would be committed to photovoltaic development and only 30% of this to solar panels. The panels are estimated to have a solar-to-electric conversion efficiency of 10%.

The solar resource is dominant in the southwest. Based on the NREL average electrolyser efficiency of 58.8 kWh/kg of hydrogen, the solar resource on a socially responsible basis is evaluated at 717 million tonnes of hydrogen equivalent annually. Applying the Norsk electrolyser system efficiency of 53.5%, the total hydrogen equivalent potential from the solar resource would be 782 million tonnes annually.

Solar thermal resource was not included.

Biomass Resource in the United States.

Feedstocks in the NREL study included crop residues, manure, wood residues, municipal waste, and dedicated energy crops (e.g., switchgrass). Food grains were excluded, but some exclusions were difficult to assess. Applicable processing methodologies were applied.

For dry feedstocks, an overall conversion of 13.8 kg bone dry per kilogram of hydrogen was applied.

For gaseous feedstocks, primarily methane, a conversion of 2.34 kg of methane per kg of hydrogen was applied (theoretical 2 to 1 ratio). This assumes an 85% efficiency of steam reforming the methane.

Crop residues, more so than wood residues and other waste, have a high volume-to-weight ratio (low density) and are naturally spread over large land areas resulting in longer transportation distances. This can involve higher transport costs to central processing plants and may economically eliminate some of these potential resources.

Most biomass will be located in high population regions from derived municipal waste.

The NREL study, based on the average 58.8 kWh/kg of hydrogen, estimates this potential resource to have the energy equivalent of 30 million tonnes of hydrogen annually (Norsk efficiency basis 32.7 million tonnes).

The conversion of biomass to electric energy does not have the variability of wind and solar so most if not all of this electricity can be supplied to the grid when required, much in the same operating manner as thermal power plants. Conversion to hydrogen for storage and by fuel cells returning the electricity to the grid on demand would be a limited exercise, if at all.

Much of the biomass energy production, because of its siting in high population areas, does not experience the constraints of transmission line capacity limitations.

COMBINED RENEWABLE RESOURCE POTENTIAL IN THE UNITED STATES EXPRESSED IN HYDROGEN ENERGY EQUIVALENT.

Renewable Resource	Primary Regions	Based on 58.8 kWh/kg	Based on 53.5 kWh/kg
Wind Energy	Central States	273 mm tonnes/yr	300 mm tonnes/yr
Solar Energy	Southwest	717 mm tonnes/yr	782 mm tonnes/yr
Biomass Energy	High Population	30 mm tonnes/yr	33 mm tonnes/yr
TOTALS		1,020 mm tonnes/yr	1,115 mm tonnes/yr

United States Electricity Consumption in 2005 and Renewable Energies Potential

In 2005, the electricity usage in the United States was 3.816 trillion kWh. World usage was 17 trillion kWh—17×10^{12} kWh. U.S. consumption was 22% of the total world usage. A more recent figure for U.S. electricity usage is given for 2007 as 3.892 trillion kWh as an estimate. It may be noted that the second highest consumption of electricity is by China at 2.859 trillion kWh.

The quantity of hydrogen equivalent to the U.S. electricity usage is based on the energy content of hydrogen at 39.4 kWhe/kg. (kilowatt hour equivalent per kilogram of hydrogen). The efficiency of conversion through a fuel cell is calculated as follows:

>Higher Heat Value (HHV) of hydrogen is 142 MJ/kg (megajoules/kg of H_2)

>There are 3,600 joules per Watt-hour or 3.6×10^6 joules per kWh

142×10^6 Joules/kgH$_2$ / 3.6×10^6 Joules/kWh = 39.4 kWhe/kgH$_2$ "e"—equivalent)

At 50% efficiency for the fuel cell:

39.4 kWhe/kgH$_2$ × 0.50 = 19.7 kWh/kgH$_2$ of electricity (approximately 20 kWh/kg)

For the portion of hydrogen produced from renewables, which may be converted to electricity, the primary developed and proven facility is the hydrogen fuel cell; alternatively, the ICE driven generator. Efficiency of conversion from hydrogen back to electricity through the fuel cell is improving based on the extensive research in this field by many technological centers. An established efficiency of 50% is a current level. The fuel cell output is therefore in the order of 20 kWh per kg of hydrogen. The degree of compression for storage and transportation, if it is not produced at the point of use, can reduce the global efficiency by 10 to 15%.

Note: Ref. Ulf Bossel, European Fuel Cell Forum, October, 2003 stated fuel cell efficiency at 50% before parasitic losses could reduce it to 45% for electricity to the grid or 40% for vehicles. He noted some claims are for 60% efficiency which may be due to starting with the lower heat value (lhv) rather than the higher heat value (hhv).

If all electricity usage in the United States was totally from renewables first stored as hydrogen the quantity of hydrogen required annually would be:

3.816×10^{12} kWh / yr / 20 kWh/kg = 0.191×10^{12} kg / yr of hydrogen

191 billion kg/yr or 191 million tonnes/yr. *
*This would be two-thirds of the reasonably developable wind energy.

The major U.S. wind energy resources are relatively remote from the main load centers and there are two significant constraints. One is the limited capacity of existing transmission lines and the second is the variability of the resource, sometimes exceeding demand and sometimes with no output. For wind energy to be the dominant supply source of electricity (subject to solar energy costs continuing to reduce dramatically) for the United States it is necessary:

- When the wind resource is producing at a high level, but during low power demand periods, to transmit this electric energy utilizing the low usage transmission systems to demand centers where it is converted by electrolysis to hydrogen which can be stored. It is reconverted to electricity through fuel cells at the load centers when demand exists.
- When the wind resource is producing at a level, high or low, and is connected to a transmission system with adequate capacity, the electricity can be delivered directly to the grid customers without any storage conversion.
- When the wind resource is producing at a high level and the transmission system lacks sufficient capacity to carry it to the customers the excess electricity can be converted to hydrogen and stored at the generating site. Then when there is capacity available on the transmission system the hydrogen can be reconverted to electricity and sent to the grid customers.

U.S. Vehicle Fuel Consumption in 2008 and Renewable Energies Potential

In the United States in 2008, vehicle fuel consumption was reported to be 378 million gallons of gasoline per day. In 365 days, the total was 138 billion gallons. (Ref. Energy Information Administration (EIA), U.S. Government—update July 2009.)

Hydrogen fuel cell vehicles use 1 kg of H_2 compared to 2 gallons of gasoline in comparative vehicles. The heat content of 1 kg of hydrogen is about the same as that of the average quality gasoline however the efficiency of conversion of hydrogen to motive power with fuel cells is double the efficiency of that of gasoline in an internal combustion engine. The conversion efficiency from hydrogen to electricity is 50% and parasitic losses will be about 10%. Electrical losses from the fuel cell to the wheels will also be about 10%. This reduces the net efficiency from hydrogen to the wheels to 40% based on the higher heat value (see note). While some of the latest most efficient diesels may claim an efficiency of 25%, the general vehicle efficiency on gasoline today is 20%.

Note: Comparison of FCEVs (fuel-cell electric vehicles) with ICEs need also to consider the reduced pollution and GHGs from WEH2 compared to gasoline, reducing the dependence on overseas oil and virtual elimination of cost inflation on the fuel.

The equivalent quantity of hydrogen to replace gasoline for a year would be

$$138 \times 10^9 \text{ gg/yr} / 2 \text{ gg} / \text{kg H}_2 = 69 \times 10^9 \text{ kgH}_2 \text{ (69 million tonnes/yr)}$$

gg gallons of gasoline

This would be equal to 7% of the U.S. total potential renewables.

U. S. ELECTRICITY AND VEHICLE FUEL USE COMPARED WITH U. S. POTENTIAL RENEWABLE ENERGIES

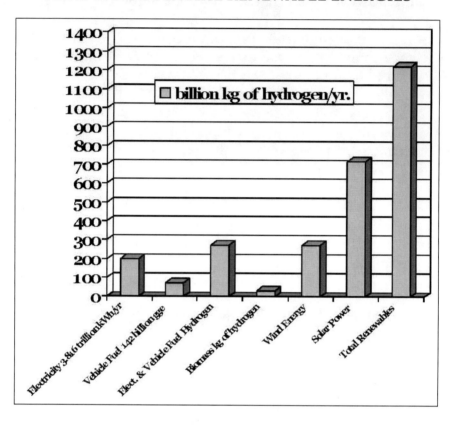

Fossil Fuels Cost for Vehicles Versus Renewable Energy Investment for Hydrogen Replacement

The United States consumes 25% of the global oil supply. The per capita consumption is six times the global average of four barrels per annum. Total world oil production in 2008 was 85.5 million barrels per day for 31 billion barrels per annum. U.S. 2008 consumption was 19.5 million barrels per day or 7.1 billion barrels per annum (23% of the world total). U.S. net petroleum imports were 11.11 million barrels per day for 4 billion barrels per annum. U.S. gasoline consumption in 2008 was 378 million gallons per day or 9 million barrels per day (42 gallons per barrel) and 3.3 billion barrels per year of gasoline. Ref: Energy Information Administration, *U.S.* Government, update July, 2009.

Fossil fuel oil now supplying gasoline to the United States at an average yield of 21 gallons of gasoline per barrel of crude oil would require on an annual basis:

378×10^6 gg/day x 365 days/yr / 21 gg/bbl of crude = 6.56×10^9 bbls/yr. of crude oil.
* gg—gallons of gasoline
* 4 billion bbls/yr of this imported bbl—barrels of crude oil

Replacement of crude oil imports of 4 billion barrels per annum by hydrogen from renewables would reduce the export of U.S. funds and reduce U.S. gasoline consumption by 60%:

Bbls. of crude per annum	Price per barrel	Total Funds Exported Annually	Total Funds Exported in Four Years
4.0 billion bbls	$75.00	$300,000,000,000	$1,200,000,000,000
4.0 billion bbls	$150.00	$600,000,000,000	$2,400,000,000,000
4.0 billion bbls.	$250.00	$1,000,000,000,000	$4,000,000,000,000

Investment Cost to Supply Hydrogen From Renewables to Replace Gasoline Now Supplied by overseas oil in the United States.

This analysis is based on supply of hydrogen from wind energy which is currently the least cost for the quantity involved, although it is expected that improved mass production and technological advancement will make solar more closely competitive. Currently about 60% of the gasoline consumed in the United States is dependent on overseas crude oil imports; the balance from domestic production. This analysis is based on replacing only 60% of the current gasoline production in the United States, sufficient to eliminate the import of foreign oil.

Total gallons of gasoline annually (2008), quantity of hydrogen for replacement, and the electricity required for electrolysis with one kilogram of hydrogen equivalent to two gallon of gasoline:

378×10^6 gg/day $\times$ 365 days/yr = 138×10^9 gallons of gasoline per year

138×10^9 gg/yr / 2gg/ kgH$_2$ = 69×10^9 kg of hydrogen annually.

69×10^9 kgH$_2$/yr $\times$ 53.5 kWh/kgH$_2$ = 3.69×10^{12} kWh/yr. (Norsk effy.)

Replacing only 60%—0.6 $\times$ 3.69×10^{12} kWh/yr = 2.2×10^{12} kWh/yr.

Wind energy installed capacity required to supply this quantity of electricity; converting to megawatts (MW), to MWh per hour, and applying a capacity factor:

2.2×10^{12} kWh/yr / 1,000 kWh/MWh / 8,760 hrs/yr / 0.30 C.F. = 840,000 MW

The capital investment includes the wind turbine generators and infrastructure together with the electrolysers. The cost currently will be in the order of $3 mm per MW installed; however, reference to NREL reports and the potential of cost improvements with mass production may reduce these costs to $1.5 mm per megawatt. The total capital investment to supply sufficient hydrogen from wind energy to replace 60% (equivalent to overseas crude oil imports) of the gasoline consumption in the United States (2008):

840,000 MW $\times$ 3×10^6/MW = 2.5×10^{12}

or at the lower projected cost:

840,000 MW $\times$ 1.5×10^6/MW = 1.26×10^{12}

Offsetting the on-going funds being exported annually for overseas oil with the one-time capital investment for the replacement by wind energy would, on a direct basis, involve the time periods indicated following, depending on the price of oil and the cost of installing wind capacity. It might be reasonable to consider the offset period to be between two and four years.

Price per Barrel of Oil	Annual funds Exported	Wind Energy Investment at Higher Capital Cost	Offset Period to Clear Capital Investment
$75.00	$300,000,000,000	$2.500,000,000,000	8 years
$150.00	$600,000,000,000	$2.500,000,000,000	4 years
$150.00	$1,000,000,000,000	$2.500,000,000,000	2.5 years
		Wind Energy Investment at Lower	
		Capital Cost	
$75.00	$300,000,000,000	$1.260,000,000,000	4 years
$150.00	$600,000,000,000	$1.260,000,000,000	2 years
$250.00	$1,000,000,000,000	$1.260,000,000,000	1.26 years

Policies and incentives which would achieve this replacement of foreign oil could result in hydrogen supply to vehicles on an ongoing basis at a fraction of the current cost of gasoline with no inflation of the energy resource.

An Analogy:

To better understand the significance of the proposed substitution of wind energy to hydrogen for gasoline from overseas oil and how to structure policies and incentives to achieve the transformation, consider this simple analogy.

A family lives in a $300,000 house and pays the landlord an annual rent which may be $75,000—or it may increase to $150,000 or much more. If the family is concerned about climate warming, then the analogy would have the house leaking formaldehyde fumes with dangerous mold in the basement. Down the street is a clean $300,000 house which can be bought and into which the family can move. They manage the financing for the clean house by an initial payment of 20% or less.

Land and Water Requirements

Land requirements for wind farm installations are minimal. Many photos show cattle grazing and grain being harvested right up to the wind turbine towers. There are spacing requirements between wind turbine generators (WTGs) in order that the disturbance of the wind leaving one WTG does not reduce the efficiency of the next receiving unit. Wind in each region generally has a primary direction, and WTGs facing that direction are generally spaced three rotor diameters apart. A second row is spaced about seven rotor diameters back from the first. This spacing will generally allow for one WTG on each 40 acres, and the WTGs may be sized at a capacity of two megawatts (1 MW per twenty acres). Foundation requirements are in the order of 30 feet by 30 feet, often in a circle rather than a square, for 900 square feet or less. A service road of twenty feet width is needed between WTG's with an average length between WTG's of one half mile for a square footage requirement per WTG of 20 ft. x 2,640 ft. = 52,800 sq. ft. Add to this the foundation requirement for a total of 53,700 square feet (1.23 acres) per 2 MW WTG or 26,850 square feet per megawatt. With 1 MW on every twenty acres (871,200 square feet), the portion of land occupied by the wind farm is 3%.

Water requirements for wind energy to hydrogen facilities are primarily for electrolysis and much of this water can be recovered in very pure form. It requires about 54 gallons per megawatt-hour. By comparison, a thermal power plant will use 300 gallons per megawatt-hour for the majority with once-through cooling; with nuclear plants water use is higher due to their lower steam temperature. For the 2005 electricity usage in the United States of 17 billion megawatt-hour the saving of water using wind and solar renewables c.f. thermal and nuclear power plants would be more than three trillion gallons annually.

Greenhouse Gas Reduction by Replacement of Gasoline with Hydrogen in Vehicles in the United States

The Global Wind Energy Council (GWEC) reports a total of 120,000 MW of global wind energy capacity at the end of 2008, saving 158 million tonnes of carbon dioxide annually. Extrapolating this greenhouse gas reduction with replacement of gasoline by hydrogen in the United States (an approximation only and may be higher), the reduction would be 1,400,000 MW / 120,000 MW x 158 M tonnes = 1,840 million tonnes annually. Note: 1,400,000 MW of WTG's if all gasoline is replaced by hydrogen.

The power sector is the largest single source of emissions, about 40% of CO_2 emissions and about 25% of overall emissions. The options for making major emissions reductions in the power sector are energy efficiency and conservation, fuel switching from coal to gas and renewable energy, primarily wind and solar power.

Within three to six months of operation, a wind turbine has offset all emissions caused by its construction to run virtually carbon free for the remainder of its 20 year life. Ref. GWEC.

Observations and Conclusions

Replacement of overseas oil and the export of funds for its purchase, which can be expected to increase in amount due to limited supply and increasing world demand, with the excessive contribution its consumption makes to greenhouse gases, strongly supports policies and incentives to hasten development of wind and solar renewable energy sources. These renewables have virtually no inflation costs and provide self-sufficiency, reduce the tensions of international commodity competition, improve the balance of trade, and divert funds going offshore to investments and jobs at home.

Enhance existing policies and incentives, which encourage research and development for more efficient and lower capital cost renewable energy systems, for greater availability of energy transmission from the productive regions to the high-demand regions, for improvement of the most promising methods of electric energy storage and reconversion and for achievement of the targeted increases in renewable energies to reduce greenhouse gas emissions.

Policies and incentives in financial terms will only require a small fraction of the vast sums exported annually for overseas oil as private investment in renewable energies need only modest encouragement. The major portion of the private investment will be found in the marketplace, earning an encouraging return.

The following chart indicates the potential for future power needs being supplied by renewables in combination with support from natural gas turbine generating plants.

United States Electricity Generation Projection Chart

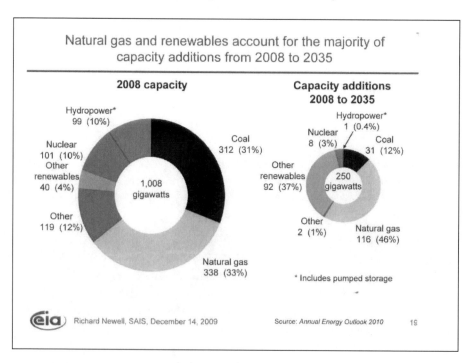

Chapter 11

WIND ENERGY TO HYDROGEN TO AMMONIA FERTILIZER

W2H2NH3

Ammonia

The common method of producing ammonia is by combining hydrogen with nitrogen in the presence of a metal catalyst. The method is known as the Haber-Bosch process.

$$3H_2 + N_2 = 2NH_3$$
$$6 + 28 = 34 \text{ (molecular weights)}$$

During very early development of hydroelectric power generation in Norway (Norsk Hydro) and in Quebec (Shawinigan Hydro (Chemical) virtually excess electricity was converted by electrolysis to hydrogen to produce ammonia fertilizer. As the market for electricity grew and natural gas became a readily available low-cost resource for hydrogen, the production of ammonia shifted to natural gas as the main resource. Current development of renewable energies and the increasing cost of natural gas has made production of hydrogen by electrolysis again attractive. Wind and solar energies also tend to be in regions which have a high demand for fertilizers, reducing transportation costs.

Ammonia is the second largest consumer of hydrogen in the world at 8.5 billion kg/yr (see chapter 2.) Some reports indicate it consumes 50% of the current world production of hydrogen. The hydrogen is virtually all produced by steam reforming methane (natural gas), and for each kilogram of hydrogen produced, there are 6.5 or more kilograms of carbon dioxide emitted to the atmosphere. The total carbon dioxide emitted annually by the ammonia production industry is therefore in the order of 55 billion kilograms (55 million tonnes/yr).

University of Minnesota Study for Ammonia Utilizing Wind Energy

A developmental study is being conducted by the University of Minnesota (U of M), West Central Research and Outreach Center reported November 16, 2007. Technology-backed participants included Norsk Hydro (now Statoil—Norway), Xcel Energy, and the U.S. National Renewable Energy Lab.

The energy source is a 1.65 MW WTG installed in 2005, producing 5.4 million kWh/year. This yield indicates a favorable capacity factor of 0.37.

5,400,000 kWh / yr / (1,650 kW x 8,760 hrs/yr) = 0.37 C.F.

A basic wind to hydrogen (by electrolysis) to storage was planned in 2006, using an ICE to convert the stored hydrogen to electricity to supply the majority of power requirements of the University's Morris campus.

The report notes that the natural gas market drives the price of ammonia as steam-reformed methane is the primary source of hydrogen for production of ammonia (refer to chapter 2).

Ammonia fertilizer was costing $100 per ton in 2002, and the cost increased in step with natural gas to over $400 per ton (Agriculture Energy Alliance, 2006), when gas increased to over $12 per million British thermal units. The project included production of ammonia from the stored hydrogen.

The project was based on hydrogen by electrolysis from the WTG with nitrogen from the atmosphere, combined in a reactor and passed through a catalyst. The reactor pressure is 800-900 psi. The W2H2NH3 pilot plant was planned with a capacity of 1 tonne/day with a cost of $3.75 million.

It is reported that the pilot plant capacity has been scaled back to 6.6 lbs. per hour of ammonia, 25 tons per year. The plant is expected to be in operation by mid-2011.

Projection by U of M for Commercial Production

The study projected a requirement of 2 GW (2,000 MW) of nameplate wind energy to produce approximately one billion pounds of nitrogen fertilizer annually (Minnesota agricultural usage). (Refer NASS 2006 MN Ag Statistics).

The study estimates the 2000 MW of WTG capital cost at $3.5 billion ($1.75 million per MW of connected capacity). Converting the electricity to hydrogen and combining it with nitrogen in ammonia production facilities had an estimated capital cost of $3.5 billion.

> 2,000 MW x 0.37 C.F. x 8,760 hrs/yr x 1,000 kW/MW = 6,480,000,000 kWh/yr.
> Hydrogen production from this wind energy
> 6,480,000,000 kWh/yr / 58.8 kWh/kgH$_2$ = 110 x 10^6 kgH$_2$
>
> Ammonia requires 6 / 34 = 0.18 lbs H$_2$ / lb NH$_3$
> 1 x 10^9 lbs NH$_3$ (1 billion lbs) requires 180 x 10^6 lbs H$_2$
> Electricity required for this hydrogen
> 180 x 10^6 lbs H$_2$ / 2.2 lbs/kg x 58.8 kWh/kgH$_2$ = 4,640 x 10^6 kWh
>
> Electricity required per lb of NH$_3$
> 4.64 kWh / lb NH$_3$ for production of hydrogen

Note: The study may be using a lower electrolyser efficiency and/or the additional wind energy is used in the process plant.

The study projects cost of production of NH$_3$ in a wide range of $0.20 to $0.60/lb with the cost of energy in the range of $0.15 to $0.36 per lb. of NH$_3$. At $0.20/lb the annual cost of the ammonia would be $200,000,000 or $400 per short ton.

> Cost of energy in the range of $0.15 to $0.36 per lb of NH$_3$
> At 4.64 kWh / lb of NH$_3$ the cost of electricity is indicated at $0.15 / lb. NH3 / 4.64 kWh / lb NH$_3$ = $0.032 per kWh.
> $0.36 / lb. NH3 / 4.64 kWh / lb NH$_3$ = $0.0776 per kWh.

If the average cost per pound of $0.20 has the average cost from the study for electricity of $0.15 deducted, then the balance of plant cost per pound is projected at $0.05.

Observations and Conclusions

- Ammonia is used directly as a fertilizer and is also combined with other elements to produce fertilizers such as ammonium nitrate.
- In remote regions with a strong wind resource and limited transmission capacity to deliver the electricity to the load centers, the excess wind energy may be converted to hydrogen and to ammonia. The ammonia can be economically and safely transported by truck or rail in liquid form to its domestic market or for export.
- A carbon credit could be available due to replacing the use of natural gas as the source of hydrogen. N plus H3 = 14 plus 3 = 17. One billion pounds of NH3 requires 3/17 x 1 billion = 176,000,000 lb. of hydrogen, or 80,000 tonnes annually. Production of this quantity of hydrogen by steam reforming methane would result in emission of 520,000 tonnes (or more depending on efficiency) of carbon dioxide annually. At a carbon credit ranging from $15.00 per tonne to $50.00 per tonne the credit would be $7,800,000 up to $26,000,000 for the 455,000 tonnes of ammonia ($17 to $57 per tonne or $15 and $52 per short ton).
- A comparison of the cost of ammonia produced from natural gas with the cost based on hydrogen produced from renewables can be made by comparing the cost of the hydrogen input in each case, with a carbon credit attached to the hydrogen produced from renewables. See chapters 2 and 3.
- Ammonia has the potential of being an energy carrier for hydrogen from renewables, with a relatively high hydrogen component carried in liquid form.

Chapter 12

MOTOR VEHICLES, EMISSIONS, and Hydrogen Fuel Cell Vehicles

Hydrogen and Vehicles

The number of vehicles (cars, trucks, and buses) worldwide has grown from 484 million in 1985 to 671 million in 1996 and is projected to 1,200 million by 2027. At the rate of growth of the number of motor vehicles in China and to a lesser degree in the other BRIC nations, the number projected for 2027 is likely to be exceeded by a considerable margin.

Hydrogen produced from renewables will find its uses decided by economics and environmental conditions. In some instances, it may best be used to produce electricity to meet peak demands, or it may find better use in vehicles. There are sufficient renewables developable on an economically, commercially, and socially acceptable basis for hydrogen to supply the world's needs for both.

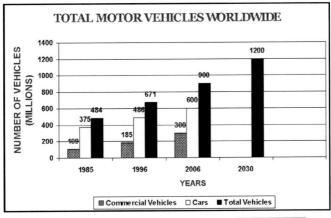

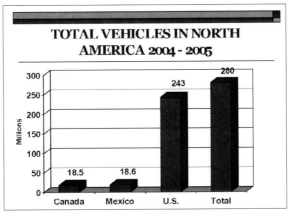

Emissions from Vehicles

Emissions from cars in 2006 averaged 180 gms/km of CO_2e (e-equivalent). Based on 20,000 km/yr/car, the *average emissions would be 3.6 tonnes / year.*

Commercial vehicles in 2006 would produce emissions in the order of 3.5 times that of cars due to larger engines, weights, and more kilometers driven.

Worldwide emissions in 2006 from 600,000,000 cars (est.) = 2.16 billion tonnes / year.

300,000,000 commercial vehicles x 3.6 tonnes/yr x 3.5 (est.) = 3.78 billion tonnes / year.

Total 5.84 billion tonnes / year.

In the United States, at a weighted *average of 7 tonnes / year of CO_2e,* the total GHG emissions *in 2004 would have been 1.7 billion tonnes.* Some authorities place this at 6.5 tonnes / year / vehicle.

In Canada, it would have been *130 million tonnes in 2006.* In 1995, Environment Canada estimated total GHG emissions from Canada road transportation sources at 123 million tonnes / year.

Greenhouse Gas Emissions (GHGs) from Internal Combustion Engine Vehicles

Continuing improvements are being made by motor vehicle manufacturers utilizing internal combustion engines (ICE's) to reduce GHGs as indicated in the following chart showing gasoline performance in 2010 compared to that of two years earlier. The improvement is in the operation mode. Total GHG emissions in 2010 were the same as for ICE's using natural gas (methane). The emissions in the operation mode for methane being less than for gasoline, but on a life-cycle basis, the GHGs reported for production of methane are higher than for gasoline.

Hydrogen produced by reforming natural gas, with its heavy component of carbon dioxide during production, has slightly higher total GHG emissions in the ICE than natural gas used directly. It shows no emissions in the operation mode but a very high component in the fuel processing stage.

Hydrogen from renewables by electrolysis is shown in the ICE to have extremely low total emissions which consist solely of fuel processing and recycling components and none in operation.

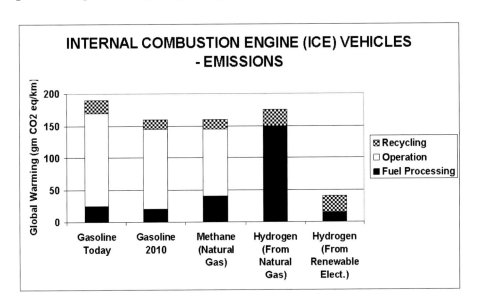

Greenhouse Gas Emissions from Fuel Cell Vehicles

The following chart provides a comparison of GHGs for fuel cell vehicles utilizing three different energy sources. Using hydrogen by reforming methane results in slightly lower emissions than when using methane directly in the fuel cell, the former has no emissions during operations but a much higher degree of emissions in the fuel processing stage.

Hydrogen produced by electrolysis from renewables has no operations component of GHGs and a very small fuel processing component (this is mainly in production of the equipment and installation of the generating system).

Vehicles fueled by hydrogen produced from renewables reduces the GHG emissions to the lowest level available with today's technology. All other components of emissions from vehicles fueled by other sources including NOX, VOX, particulates, etc. do not exist in fuel-cell vehicle exhaust.

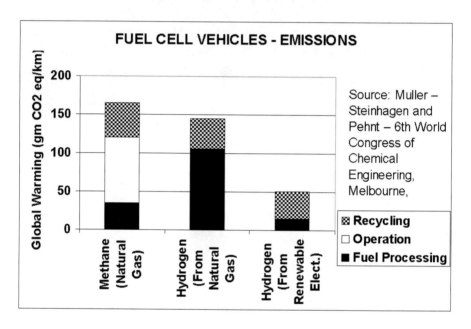

Electric Battery Fueled Vehicles

Governments are encouraging the manufacture and use of "plug-in" vehicles. Several countries have set early targets of one million or more on the road within the next several years.

Battery size, capacity, and other key characteristics are being improved at a rapid rate as manufacturers rush to supply competitive models to the market. Cost is currently higher than for conventional and hybrid gasoline-battery vehicles which compete in terms of fuel consumption and low emissions. Replacement of used batteries is a significant cost so greater lifetime is also critical.

Battery charging is generally done at night when the vehicle is not in use. This is also favorable for the utility as this is their low-demand period and excess generating, transmission, and distribution facilities are available at only the variable costs for them. Multirate schedules provided by the utilities will encourage use of electricity during low-demand periods. Some utilities have a multitiered rate schedule which will be a discouragement to electric vehicle operators as the customer with increased usage pays a higher rate for the additional component (a system which was designed to encourage users to save on their electricity consumption). An example of the "discouragement" rate schedule is reported in one region of the United States as a basic $0.12 per kWh with higher consumption at up to $0.30 per kWh.

Current EV models have an electricity consumption (battery charging) over a wide range, with the average about 0.15 kWh per km or 0.24 kWh per mile.

Battery plug-in vehicles which claim zero emissions do not consider the GHG emissions created by the source of their electricity supply. In the United States and most of Asia, the major component of electricity supply is coal fired, which has the highest level of GHG emissions. Coal-fired thermal plants also operate best at full load, and low nighttime loads are undesirable both financially and in terms of operating factors. Coal-fired plants will be the major supplier of the electricity used to recharge the batteries, at least until renewable energy sources are more developed. The EPA (U.S.) has reported the average CO2 emissions from the United States mix of power generation are 1.363 lb per kWh. For current EV models with 0.15 kWh/km, the emissions per the EPA report would be

1.363 lbs/kWh x 0.15 kWh/km = 0.2 lb s/km.

At an average annual driving distance of 20,000 km/yr the GHG emissions associated with an EV would be

0.2 lbs. CO_2e/ km x 20,000 km/yr = 4,000 lbs/yr or 2 tons.

A target of one million EVs would be responsible for 2 million tons of GHGs annually. The emissions would probably be higher than this due to coal-fired plants possibly offering the lowest off-peak rates because of their operating characteristics.

Preferential use of off-peak wind-powered electricity would virtually eliminate the GHG emissions for which EVs would otherwise be responsible.

Compressed Air Vehicles (CAVs)

Vehicles driven by compressed air have been developed to a degree and in operation for a century. One of the earliest applications was their use in motor-driven equipment used in underground mines. They were safe because there was no sparking or other means of igniting gas in the mine.

The desire to reduce GHG emissions from vehicles has drawn attention again to compressed air vehicles. In order to achieve an acceptable range competitive with the hybrid battery-gasoline vehicle, a similar combination may be needed. The overall efficiency may not be as good, and when the emissions from power generation used to compress the air is accounted for, the emissions may be the same or even greater than with today's hybrid.

Observations and Conclusions

- Motor vehicles are the second largest emitter of GHGs.
- Government-imposed standards on gasoline and diesel—fueled vehicles are significantly reducing the GHG emissions; however, they will remain a major contributor.
- Electric battery vehicles will reduce GHGs relative to gasoline fueled units, but due to the current mix of electricity supply, they will be responsible for about 60% to 70% of the GHGs of gasoline or diesel vehicles
- Electric battery vehicles which are charged from renewable energy will have no emissions, except life-cycle manufacturing and recycling.

- Motor vehicles which are fueled by hydrogen from renewable energies will have almost no GHGs, except the minor amount related to the manufacture and recycling of equipment and components. Replacement of gasoline and diesel by hydrogen from renewables will be sufficient to significantly reverse the accumulation of GHGs in the atmosphere.
- Hydrogen from renewables for vehicles will eliminate the need for overseas oil.

Chapter 13

HYDROGEN FUEL CELL
OR
ELECTRIC VEHICLE SUBSTITUTION
FOR
GASOLINE AND DIESEL

World Requirement for Hydrogen Vehicle Fuel by 2050

This projection is based on all vehicles being converted to fuel-cell operation by 2050. It is based on a total of 900 million vehicles using 360 billion gallons of gasoline (gg) in 2006 (400 gg/vehicle), rising to 1,500 million in 2050. The pro-rata consumption of gasoline in 2050 would be 600 billion gg. Replacing this by hydrogen, currently at a rate of 1 kg H_2 /2 gg but assuming an improvement of efficiency of 10%, the conversion rate would be 1 kg H_2/2.2 gg. The 600 billion gg requirement in 2050 would be 270 billion kilograms of H_2. Producing this hydrogen from electricity by electrolysis at today's conversion of 58 kWh/kg H_2 would require

> 270×10^9 kg H_2 × 58 kWh /kgH_2 = 15.7×10^{15} watts or 15.7 petawatts.
> Partial Ref. P. Kruger: World Nuclear Association symposium 2004.

Gasoline Consumption in North America and Hydrogen Substitution

Cars, pick-ups, and light trucks driven 20,000 km/yr (25 mpg, 40 km/g, 16 km/l—U.S. gallons) average 500 U.S. gallons per year each.

One U.S. gallon is equivalent to 1 kg of hydrogen in terms of energy content. However, due to the higher efficiency of conversion of hydrogen to motive power compared to gasoline in the internal combustion engine, one kilogram of hydrogen will move a vehicle more than twice the distance compared to one gallon of gasoline.

In 2005, there were 280 million motor vehicles in North America. Approximately 90 million would be commercial vehicles consuming mainly diesel, with equivalent gasoline consumption about three times that of light vehicles due to larger engines, weights, and more kilometers driven.

Total hydrogen required to fuel the light vehicles in North America would be

$$(280-90) \times 10^6 \times 500 \text{ gg} / 2.2 \text{ gg} / \text{kgH}_2 = 43 \text{ billion kg/yr.}$$

This is equal to the present total world production of hydrogen.

To fuel the trucks and buses would require 120 billion kilograms of hydrogen annually. This is between 10 and 12% of the reasonably available potential of renewables in the United States.

The total for all vehicles in North America would be approximately 160 billion kilograms of hydrogen annually or four times the present world production of hydrogen.

Gasoline and diesel consumption would be reduced by over 300 billion gallons annually or 12 to 14 billion barrels of crude oil annually; this is almost double the amount of crude oil now imported into the U.S.

Electric Vehicles (EVs) in North America and Reduction of Gasoline Consumption

The GM Volt electric vehicle claims an operating cost with electricity of 2 cents per mile (equivalent to 70 cents per gallon of gasoline with a vehicle achieving 35 miles per gallon) compared to 12 cents per mile for a gasoline-fueled vehicle with gasoline at $3.60 per gallon (i.e., operating at 30 miles per gallon). This does provide one commercial unit's indication of gasoline saving by an electric vehicle. With a consumption of electricity at an average reported 0.15 kWh/km or 0.24 kWh/mile, and the total cost of the electricity per mile is 2 cents, cost in this case is 8.3 cents per kWh for the electricity.

MOTIVATION FOR CONVERSION TO ELECTRIC VEHICLES

A number of countries, including the United States, have each committed to a target of 1,000,000 electric vehicles within the next five years.

A competitive EV now in production using a lithium-ion battery can travel 400 km or 250 miles per battery charge with an equivalent fuel efficiency of 1.74 liters/100 km (135 mpg).

One EV of this type, driving an average 20,000 km/year (12,430 miles/year), would displace gasoline consumption based on an ICE at 25 mpg of approximately 500 gallons per year. One million EVs would displace gasoline consumption of 500 million gallons/year. Refining one barrel of crude oil produces 21 gallons of gasoline. The introduction of one million EVs would reduce U.S. crude oil requirement by

1×10^6 vehicles x 500 gg/yr / vehicle / (21 gg/bbl. Crude) = 24 million barrels annually.

The total vehicles in the United States in 2005 was 243 million, of which approximately 120 million were automobiles. Replacing one million ICE's (automobiles) by EVs represents about 0.8% of the total automobiles in the United States.

The EV's in production or at the final development stage have a consumption range of electricity of 0.1 to 0.2 kWh per km or 0.16 to 0.32 kWh/mile. Twenty percent of this power is lost in battery efficiency.

Assuming an EV's average achievable consumption of 0.15 kWh/km and 0.24 kWh/mile, the annual electricity consumption of one million EV automobiles driving 20,000 km/year would be 20 billion km x 0.15 kWh/km = 3 billion kWh or 3,000,000 MWh/year.

For this electricity to be supplied by wind energy, in order to virtually eliminate the greenhouse gas emissions if supplied by coal-fired plants, would require 3,000,000 MWh/8760 hrs/yr = approx. 3,400 MW. The installed wind energy at 30% C.F. would be 11,300 MW (approximately 50% more than the amount installed in 2008). At $2 million per MW, the capital investment would be $22.6 billion. The direct wind energy could supply the EV requirements if it was a steady supply, primarily at night when most EVs would be charging. Balancing the wind energy to the EV requirements would require exchange of power from wind with other power sources on the grid and/or the ability to store the wind energy and deliver it when the EVs are in charging mode. Cost of the wind energy may be in the order of 8 cents per kWh, and at an average 0.24 kWh/mile, the equivalent fuel cost would be 1.92 cents/mile. Gasoline in an ICE at 25 mpg and $2.50 per gallon would cost 10 cents per mile.

A conventional gasoline-powered automobile emits 250 grams of CO_2/km. One million ICEs replaced by EVs driving 20,000 km/year with the battery-charging electricity for the EVs supplied by wind energy would reduce atmospheric greenhouse gas emissions by 20,000,000,000 x 0.25 kg = 5,000,000,000 kg/year or 5,000,000 tonnes/year.

EVs, generally driven mainly in daytime, require battery charging during nighttime hours. This provides an advantageous feature with utilization of wind energy, which can feed directly into the grid during daytime and at night, when grid demand is lower, can be used to charge EV batteries.

EVs can utilize solar power, which is increasingly being installed in residences, by utilizing a second battery. This can be charged during the day and exchanged with the spent battery for next day's use. This feature requires consideration of the weight and ease of replacement of the batteries.

Electric Vehicles Supplied from the Grid with Current Electricity Supply

The electric vehicles coming on the market will generally be plugged into a power supply from the regional grid. The majority of electricity supply in North America, China, and many other regions is from coal-fired power plants. Coal-fired plants operate most efficiently at full load, and when the grid they are supplying has a low nighttime demand, they are curtailed. Their fixed costs of investment, staff, annual maintenance, and G and A continue twenty-four hours a day; and only the resource supply of coal is the variable. Any price they can obtain during low-demand periods which is greater than the nominal cost of the coal is to their advantage. This source of power will therefore predominate in the supply of battery recharging at night for electric vehicles.

Even in regions with a relatively abundant supply of hydroelectric power, those utilities import very low-priced coal-fired power during low-demand periods while storing their own supply behind their dams to be used during high-demand, high-value periods. Only renewable energy sources which are supplying during off-peak periods will be selectively chosen to provide battery charging to the extent that the renewable energy is available.

Conclusions

- Electric vehicles with batteries charged by renewables are responsible for an extremely low level of greenhouse gas emissions. They do have some emissions charged to them for the manufacture of the vehicle and for its ultimate disposal, also varying degrees of concern regarding the battery.
- Where the current electrical supply to the EV battery will be from coal, the emissions from that source must be addressed. It will need to be compared to the emissions from the production of gasoline from crude oil and the emissions from the gasoline in the internal combustion engine. As described earlier EV vehicles with the current mix of electricity generation will be responsible for a level of emissions 60 to 70% of those from gasoline fueled vehicles.
- **For government policy and incentives provided to enhance the manufacture and use of EVs, it is necessary at the same time that they encourage the production and use of emissions-free renewable energy.**

Chapter 14

ALTERNATIVE ENERGY STORAGE OF VARIABLE RENEWABLES

Storage of Electric Energy:

Storing electrical energy in an efficient manner with rapid recovery to the grid is essential to the significant development of nonpolluting and limitless renewable energy. In most of the developed countries, where renewable energy is highly promoted, there is a lack of policies and incentives for optimizing the best methods of storing energy. Storing energy is not new to humanity. It started when humans began to use fire. Since then, they have stored wood and other biomass for heating and cooking. They have stockpiled coal and cattle dung. These are the first examples of energy storage.

For centuries, water wheels on streams drove grain grinders. Dams were built to assure their operation on a year-round basis. With the advent of electricity and water-turbine generators, huge dams were built to provide reliable supply with both energy and capacity improved by storage.

Thermal power generating plants of today stockpile coal as a precaution against interruption of supply. Crude oil and gasoline are stockpiled in vast quantities against possible interruption of supply on an international scale. Natural gas pipelines store huge quantities of gas in underground caverns to use as backup to limited pipeline capacity when demand is high and in case of pipeline failure.

The investment in these various forms of energy storage for the facilities and for the products is measured in hundreds of trillions of dollars. Development of renewables to their full potential similarly requires practical, proven, and efficient storage.

Renewables which are variable due to nature can have their energy stored by various methods for later use. These include electrolytic batteries, flywheels, compressed air, pumped hydro, molten salt, electrostatic capacitors, magnetic energy storage, ammonia, stored hydro, and hydrogen.

Storage of electric energy is achieved in most systems by converting it from kinetic to potential energy. Some form of storage is essential as the proportion of variable renewable electric power grows in order to "firm" the supply to meet the variable demand on the grid.

Grid scale electric power storage is a key need for rapidly diversifying electricity systems when wind energy exceeds twenty to 30% of the electrical load. (Ref. EPRI and DOE.)

The storage can be used to firm variable renewables and also of great importance financially; it can be used to transmit conventional thermal, nuclear, and stored hydro power over low-demand lightly loaded transmission and distribution systems with the stored power returned to the grid during peak-demand periods when the transmission and distribution systems are overloaded. New transmission and distribution capacity, which encounters major NIMBY opposition, and its cost is avoided. New generating and spinning reserves for peak-demand periods and unit failures are not needed.

An example of the need to develop economical and responsive energy storage is illustrated by difficulties being experienced in Germany, where wind energy now represents 7.5% of the total generation. In Germany and also in Spain, there are occasions at nighttime when the grid has a surplus of power and it is difficult to turn down the wind. The utilities, on some of these occasions, have a negative price for the electricity. This clearly shows the importance of energy storage and its economic value.

Batteries appear to offer a practical and economic alternative for fast response but with limited capacity and fairly high cost.

Stored hydro exists in considerable magnitude, is limited by terrain and climate, and is already largely optimized except for opportunities which still exist in some less developed regions. China has major stored hydro developments in progress. It does offer significant synergism for variable renewables, and this storage—with good response—will add considerable value to the resource where it is grid-connected to renewables. Its capability is limited when the dams are full.

Compressed air energy storage may be one of the best systems for energy storage, with its high capacity potential, long storage periods with little loss, and relatively fast response of ten to fifteen minutes to full load when used in conjunction with natural gas turbines. Efficiency and costs appear favorable.

Quick-response storage to firm up renewables energy can improve the value of this energy source and remove some of the risk of liabilities which may otherwise be imposed by the buyer-utility. Forward dispatch predictions on a firm supply basis can be made. Providing quick response from stored energy can involve more than one system. One system of smaller capacity may provide virtually immediate response integrated with a system of large capacity but slower response.

Utilities make provision for the unpredictability of variable renewables energy by holding generation reserves. This reduces the real net value of renewables energy and emphasizes the importance of providing energy storage.

Renewables energy storage with rapid response provides several operational/financial benefits:

- Utilities operating reserve offset, reducing required investment
- May be used to reduce peak demands from the utilities systems which can defer or eliminate the need for new transmission, distribution, and generating facilities to serve peak-demand periods.
- Black start
- Eliminate the need for emergency generating systems for critical facilities.

The alternative methods of storage need evaluation compared with hydrogen energy storage in terms of response time, investment, life-cycle costs, performance, efficiency, limits of capacity, limits for short-term or long-term storage, operational and maintenance costs, disposal costs, and greenhouse gas emissions.

As the proportion of variable renewables increases to 20% or more of a utility's total sources of supply, the operational regulation requirements become more significant and costs are incurred. It is important to the continuing growth of renewables that storage and its benefits be evaluated for the entire system.

While fast-response gas turbine generators are currently the main source of new power supply, the major existing generation is from coal-fired and nuclear plants, which are unsuitable for fast response and lose efficiency along with increased greenhouse gas emissions (in the case of coal-fired plants) when load is reduced (during low-demand periods).

Need for Policies and Incentives

The electric system operators (ESOs) which manage large grid systems and collaborative integration with semicontinental electric systems must play a leading role in development of fast-response limited energy (FRLEs) storage systems and energy storage firming systems (ESFSs). FRELs provide response in seconds to variability of supply from renewables and of demand. ESFSs provide large-capacity energy storage for arbitrage of low-cost off-peak power with high-cost peak power. These two types of storage systems may be developed by existing utility operators or by independent energy storage investors.

BATTERY STORAGE—UTILITY SCALE

There are a number of types of batteries experiencing rapid development to improve their utility scale application. One major advantage of this form of energy storage is rapid response, a highly valuable characteristic for a utility system operation. These are in the category of FRLEs.

Some of the more advanced types of batteries for utility service are:

Lead-acid batteries, one of the oldest forms of energy storage systems with up to 40 MWh in utility service. Several developers have systems over one megawatt-hour. These batteries have a short life cycle, and the amount of electricity they can deliver is dependent on the rate of discharge; fast discharge means less quantity.

Lithium-ion batteries have the advantages of high-energy density, over 300 kWh per cubic meter, very high efficiency near 100%, life cycles over 3,000. The disadvantages are the need for overcharge protection and special packaging. Cost has been high at over $600 per kilowatt-hour. This type of battery is finding common use in electric vehicles.

Metal-air batteries are compact and low cost. However, difficulties include rechargability with low cycles and a low efficiency of about 50%.

Sodium sulphur (NAS) batteries operate at 300°C with an efficiency of about 90%. In the order of 300 MW are in operation in Japan with six-hour peak shaving ability. A large Japanese installation of over 30 MW and 245 MWh is used for stabilization with a wind farm. About 9 MW is used in the United States for peak shaving.

Vanadium Redox batteries have supplied relatively large energy storage systems utilizing tank-stored electrolyte.

Zinc bromide (ZnBr) batteries are also under improved development.

Sorne Hill, Ireland

A system is being planned for 2 MW for 6 hours with installation of a VRB-ESS at a cost of $9.4 mm. It will have the ability to provide 3 MW for 10-minute periods. The energy source is 32 MW of wind turbines. VRB Power Systems Inc. report a global efficiency of 70%.

An electrical system and financial modeling study by Tadbury Management Limited indicates the battery storage facility would generate an IRR of 11.7 % (pretax). Commercial structuring of the investment would generate an IRR of about 17.5% (after tax) for an equity investor.

Luverne, Minnesota

Xcel Energy (Northern States Power Co.) and Minwind Energy have combined to develop a 1 MW sodium sulfur battery installation, which will be integrated with Minwind's 11.5 MW wind farm.

Palmdale, California, Water District

The water district is developing a 450 kW zinc bromine battery system to provide backup and peaking supply for their 950 kW WTG, 200kW natural gas engine and 250 kW water turbine generator. The technology is reported to have the potential to relieve transmission and distribution capacity and reduce the potential for blackouts.

King Island, Hydro Tasmania, Australia

King Island has provided some of its energy from wind turbines since 1998, replacing a portion of its normal supply from diesel generators. In 2003, more wind energy was added along with a Vanadium Redox battery of 200 kW capacity. The battery offsets to some degree the variations in the wind energy supply.

Flywheel Energy Storage

The energy storage is kinetic to kinetic and back to kinetic. These systems are also in the category of FRLEs.

Tehachapi, California, Flywheel Energy Storage

Beacon Power's flywheel energy storage is intended to improve frequency variation in this leading wind farm region. It is considered possible that this type of energy storage system may reduce peak demand on transmission systems.

Study Indicates Energy Storage Should Be an Integrated Grid Solution Southern California Edison CAISO Renewable Integration Study

It is suggested that storage of variable renewable energy be considered part of the grid facility rather than part of a renewable energy generation system.

It is observed that fast response of storage reduces the forecasting error from variable renewables.

Storage ability to meet capacity, fast response, grid, and VAR requirements from variables should earn compensation as an independent FRLE facility.

COMPRESSED AIR ENERGY STORAGE (CAES)

Off-peak or variable renewables power is used to drive turbo-compressors pumping air at about 1,000 psi into sealed caverns. This form of energy storage has very high capacity and can be stored for long periods of time with very little loss of efficiency. It is in the category of an energy storage firming system (ESFS). Cost of storage is typically in the order of $500 per kW but varies depending on the type of storage cavern available.

On demand, the compressed air is mixed with natural gas in a gas turbine generator to supply electricity to the grid. The gas turbine normally provides two-thirds of its energy to compress air from the atmosphere to mix with the natural gas entering the turbine. Compressed air from storage provides this two-thirds, so the natural gas required is reduced to 33% for the same electrical output. Greenhouse gas emissions are reduced accordingly, net of GHGs which may have been used to store the compressed air; none if from renewables.

CAES is a means of storing very large quantities of energy for short or long periods of time with very good performance efficiency reported in the range of 70 to 85%. The loss is primarily in the heat from the compressor, which is generally not recoverable.

The air is compressed by utilizing renewable energy from wind or solar systems or from low-value off-peak conventional power generation facilities. When utility system demand is high, the compressed air is returned to feed into a gas turbine generator, mixing it as the combustion air with the natural gas entering the turbine. The gas turbine exhaust is used to preheat the compressed air being drawn from the cavern. A conventional gas turbine generating plant has a large air compressor upstream of the turbine on the same shaft, and two-thirds of the energy in the natural gas is used in the turbine to drive the air compressor. When the combustion air is from the CAES, 100% of the natural gas entering the turbine is used to produce electricity. If the CAES is from renewables, then the electricity so generated releases 66% less greenhouse gases than a conventional gas turbine generating system.

Huntorf, Germany, Compressed Air Energy Storage

This facility is 290 MW built in 1978 for E. N. Kraftwerke. The system has a response time of several minutes. During low-cost, low-demand periods, electricity from base-loaded plants or from excess variable renewables drives compressors which store compressed air under high pressure in underground salt strata caverns with over ten million cubic feet of capacity. During peak load periods, the compressed air supplies the compressed combustion air section of a gas turbine generator. In a normal gas turbine generator system, two-thirds of the energy produced by the gas turbine is used to compress air from the atmosphere to mix with the gas in the turbine. With compressed air from underground storage, the full energy of the gas turbine is used to drive the generator, recovering the energy used to compress the air to storage except for the compressor-storage efficiency loss.

The compressor operation uses 60 MW over a 12-hour period, and the gas turbine produces 290 MW over a 3-hour period.

- The CAES facility provides a quick response in case of failure of another power plant, giving time to bring a cold plant on line.
- It offers an alternative to purchasing expensive peak-demand period power.
- It provides peak shaving.
- It provides quick response for changes in wind energy supply.
- It has black start capability.

Efficiency: Input to compressed air 60 MW x 12 hours = 720 MWh.

> Gas turbine output 290 MW x 3 hours = 870 MWh. Compressed air provides 2/3 of energy requirement = 2/3 x 870 MWh = 580 MWhe. Efficiency of storage system 580/720 = 80%.

The Huntorf CAES system uses the generator as a motor to drive two compressors, a low-pressure unit and a high pressure one, when converting and storing the electrical energy as compressed air.

McIntosh, Alabama, Compressed Air Energy Storage (CAES)

This facility, costing $65 mm, was started up in 1991. Air is compressed and stored in a salt cavern of twenty million cubic feet at 1,250 psi when low-cost, low-demand electric power is available. During peak-demand periods, the compressed air is supplied to a gas turbine generator, displacing the need for the turbine to compress its combustion gas and increasing the electric generating capacity of the turbine by almost three times. A single train Dresser-Rand turbine expander/compressor is utilized. The

110 MW plant is owned and operated by the Alabama Electric Cooperative. The plant can reach full load in less than fifteen minutes with 2,600 MW/hr.

CAES Projects in the Development Stage

There are seven CAES projects in the planning and development stage in the United States. There are two in Germany and one in Ireland.

Kern County, California (CAES)

Pacific Gas and Electric are proceeding with a $355-million CAES facility which will pump the compressed air into an aquifer storage system.

Watkins Glen, New York (CAES)

A 150 MW system is being developed with the compressed air stored in a salt mine. The investment cost is indicated to be $125 million.

Norton Energy, Ohio (CAES)

This facility is planned for 2,700 MW with an energy storage potential of 43,000 MWh. The compressed air will be stored in a limestone cavern at 1,500 psia.

Matagorda County, Texas (CAES)

Ridge Energy Services is planning a system with four 135 MW Dresser-Rand units. The air will be stored in salt dome caverns.

Justification for CAES Systems

Increasing wind energy (variable renewable) is now the driving force for CAES projects. The economy of reducing peak demand and eliminating the cost of generation, transmission, and distribution to serve only the peak demand can justify CAES systems but does not appear to have been evaluated on that basis by utilities.

CAES Systems Benefits

- Low environmental impact
- High reliability
- Lowest cost
- Utility scale
- Utilizes existing generation, transmission, and distribution during off-peak periods, eliminating the need for multibillion expansions of those facilities.

A Case Study Evaluating Compressed Air Energy Storage

NO NEED FOR TWO 500 KV D.C. TRANSMISSION LINES FROM EDMONTON TO CALGARY

There are three practical and economical alternatives which separately or in combination will be more acceptable to the public and provide more generated power instead of a loss of power on transmission lines.

Each alternative utilizes the storage of low-cost off-peak power in the form of compressed air in subterranean caverns with no hidden hazards. The basic system is referred to as the compressed air energy system (CAES). Such systems have been used for decades in Europe and the United States. Electrical energy can be converted by air compressors to high pressure air stored underground. This is converting kinetic energy into potential energy. When the compressed air is returned through gas turbine generators to feed electric power into the grid, the energy loss is in the order of 20%, a system efficiency of about 80%. However, the power returned to the grid is worth five to ten times that of the energy used to store it.

Alberta, including Southern Alberta, has a wealth of subterranean storage caverns.

The typical gas turbine generator has three components on a single shaft. The physically largest component is an air compressor which draws in ambient air and compresses it to mix with high-pressure natural gas, both then entering the second component—the turbine—which is similar to the turbines used on aircraft. The gas in combustion expands, driving the turbine and the third component, the electrical generator. Over 60% of the energy in the natural gas being supplied to the turbine is required to drive the air compressor. The CAES supplies the combustion air to mix directly with the natural gas entering the turbine, eliminating the need for the air compressor on the turbine shaft. The output of the turbine generator is now more than doubled.

CAES from Existing Power Generation:

Coal-fired power plants in Alberta operate at very low output levels during nighttime, an undesirable performance. They do sell some of this off-peak power through an intertie transmission line to B.C. Hydro. B.C. Hydro purchases this power at a price only slightly more than the variable cost to the coal-fired plants, which is primarily the low cost of their fuel—coal. B.C. Hydro stores the equal amount of this imported power by holding the water back behind their dams. Then during daytime power demand peaks, they sell some of this power back to the Alberta grid at five to ten times the price which they had paid for it.

The coal-fired excess nighttime off-peak power in Alberta can be stored in a CAES system in Southern Alberta. The power is transmitted from the Edmonton region to the Calgary area over the existing transmission lines which have excess idle capacity at that time. Then during daytime peaks, the CAES supplies stored energy to gas turbine generators in the Calgary region. There is a 20% loss, but the electric power is converted from a low-demand period extremely low price to high-value peak demand power. The need for more transmission capacity from Edmonton to Calgary is eliminated by utilizing the low-use nighttime capacity of the existing lines. The investment cost for this alternative is less than the cost of the proposed new 500 kV transmission lines.

In addition, there is the direct energy profit through buying low-value power and converting it to high-value power.

CAES With New Combined Cycle Gas Turbine Generators (CCGT):

Sufficient CCGTs may be installed in Southern Alberta to supply the peak demand in the Calgary region and eliminate the need for the new 500 kV transmission lines from the Edmonton area. The price of natural gas had been a significant concern; however, the vast reserves resulting from new drilling technology ensure a reasonable cost for some time into the future.

These CCGTs can be *operated at full capacity full time* with the off-peak power stored as CAES. The CAES-supplied gas turbine generators will supply the peak loads which reduces the requirement for complete CCGTs. The capital investment in gas turbine generators is only a fraction of that for all other forms of power generation, and the lead time for installation is short. Carbon dioxide (greenhouse gas) emissions are only 50% of that of coal-fired plants. The installed cost of one thousand megawatts of CCGT's at $600,000 per megawatt would be $600 million dollars, only a fraction of the cost of the 500 kV transmission lines; and they would be supplying product—electric power.

CAES With Wind Turbine Generators

Southern Alberta has proven an outstanding wind energy resource with considerable development to date. There are projections of the potential of wind energy development in Southern Alberta in the order of ten thousand megawatts. Alberta's current total power generation is about twelve thousand megawatts. There is concern for the system operations with major wind energy development due to its variable nature. For this reason, truly major development will be more readily accepted when it is combined with an energy storage system.

When the wind energy production is high, simultaneous with a high-grid demand, then it may be fed directly into the system.

During low-demand periods, the wind energy will be excess, and its real value will only be captured by storing it. This can be done by using it to drive compressors, storing the energy in a CAES. The CAES then operates with a gas turbine generator system to supply the grid when demand is high and the wind energy source is low.

The CAES with wind energy supplies its component of power through gas turbine generators with virtually no greenhouse gas emissions. CCGT power plants have a much greater ability to alter their output to match wind power supply than any other thermal power plant.

The Wind Energy system, once paid for, will produce power long into the future with exceptionally low cost per megawatt-hour as there is no energy resource cost component. It will produce the electric power with almost no greenhouse gases. The alternative of the new transmission lines will produce no new product but will drain 6 to 8% of the greenhouse gas-producing power it transmits.

PUMPED HYDRO ENERGY STORAGE (PHES)

Two water reservoirs are utilized with a difference of elevation to provide the hydraulic head. During daytime peak, electricity demand water from the upper reservoir is released through a penstock (pipe), discharging through a water turbine driving an electric generator, to the lower reservoir. During off-peak hours, the generator acts as a motor, drawing electricity from the grid and driving the turbine in reverse as a pump which lifts the water up through the penstock, delivering it back to the higher reservoir for reuse.

These systems are in the category of energy storage firming systems (ESFS's).

Some hydro storage projects with high dams can be adapted to pump water back during low-demand periods.

Underground pumped storage using caverns or abandoned mines can be used.

High-elevation water reservoirs which discharge to the ocean, pumping the saltwater back during low-demand periods, are another potential application.

Efficiencies are in the range of 70 to 85%.

There are over 90,000 MW of pumped storage in operation, about 3% of the total world electricity generation.

Capital costs are generally high; however, operating costs, maintenance, and the inflation factor are low.

Thissavros, Northern Greece, Pumped Hydro Storage

This project was completed in 1996 with three 100 MW reversible pump-turbines on the Nestos River.

TVA Raccoon Mountain Pumped Storage

This Tennessee Valley Authority-pumped hydro storage in South East Tennessee has a capacity of 1,600 MW. Reversible turbine-generators draw down water from the upper reservoir during peak-demand periods, and the same equipment pumps the water back from the lower reservoir during off-peak periods. Efficiency is between 70 and 85%.

Vianden Hydro, Luxembourg

This pumped storage was completed in 1964 with 10 generators for a total capacity of 1,100 MW. It has a global efficiency of 74% with an annual production of 1,650 GWh.

Tianhuangping, West of Shanghai

This pumped storage was completed in 2001 with a capacity of 1,800 MW. It has a head of nearly 2,000 feet, with a reservoir capacity of eight million cubic meters. It has an overall cycle efficiency of 70%.

Oglethorpe Power, Rocky Mountain

This pumped hydro facility with a capacity of 848 MW was completed in 1995. It has a 1,000-acre reservoir. The output can reach full capacity in fifteen minutes.

It pumps the water to the upper storage in eight and one-half hours using 7,500 MWh and can draw down 6,400 MWh over seven and one-half hours. This indicates an efficiency of 85%.

MOLTEN SALT ENERGY STORAGE

Andasol Solar Power Station, Granada, Spain

This is Europe's first parabolic trough commercial power plant, 50 MWe, started up in November 2008. Part of the solar heat produced during the day is stored in a molten salt mixture and used to generate electricity during the night. A steam turbine generator produces the electricity.

Solar Power Tower Plant of Solar Tres, Spain

This 15 MWe solar-only power plant of SENER Utility uses a United States 16-hour capacity molten salt technology for energy storage. It can deliver electricity 24 hours per day from a variable source.

ANHYDROUS AMMONIA ENERGY STORAGE

Ref: Leighty Fdn. NHA-March, 2009.

Ammonia (NH^3) is a carbon-free fuel with a 17.8% mass hydrogen content, a high-energy density as a low-pressure liquid, suitable for storage in carbon-steel tanks with no embrittlement. This could provide, from wind and solar variables, *annually firm* energy storage.

A solid-state ammonia-synthesis (SSAS) technique may be applicable.

The stored ammonia may be reformed to hydrogen and supplied to a fuel cell to produce the electric power. Alternatively, it may be used in an ICE vehicle or provide heat through a space heater.

As an example, one ton of hydrogen would be stored in 6.5 tons of ammonia, in a 10,000-liter steel tank costing about $25,000.

This alternative method of storing hydrogen energy may require research and development. If proven successful, it could become a significant alternative to compressed hydrogen storage and transmission. It could become an excellent system for year-round reliable supply of electric power and other energy uses from nature's variable renewables.

Refer also to chapter 12.

ELECTRIC VEHICLE ENERGY STORAGE

The world is moving rapidly forward with development and commercialization of electric vehicles which are being designed to provide as nearly as practical the likeness and features of the gasoline and diesel engineered cars and light trucks. The primary motivation is reduction of greenhouse gasses affecting climate change and direct pollution of the air people breath.

It is probable that the pattern of use for electric vehicles will see them used during the day for commuting and delivery service, with the batteries recharged during the night. Many, possibly most, will be charged by plug-in to a residential circuit.

Battery development resulting from phenomenal research has achieved huge improvements in storage capacity, recharging time, and weight. Driving ranges up to 400 km are available in some models. Safety, with respect to the batteries, continues to be of some concern, as well as disposal.

The electricity demand by electric vehicles will primarily be during normal low-demand night time, which means no increase in generating capacity, transmission, or distribution. Utilities should be encouraging this by metering low-demand periods at a significantly lower rate per kilowatt-hour using the newly developed smart meters.

This form of energy storage can make available battery-stored power for feed into the grid during higher daytime demand periods. Each vehicle may be capable of supplying in the order of thirty to forty kilowatt-hours each day to the grid, several times the consumption of the average household.

It will reduce emissions by a net of the gasoline and diesel vehicle emissions less the emissions of the electric generating stations which produce the power for the batteries. Emissions from the power generating stations supplying the electricity for charging the vehicle batteries will vary country by country. In the United States electric power is supplied 55% from coal, 9% from natural gas, 4% from oil, and the remaining 32% from nuclear and renewables. In France, the majority of their power is from nuclear plants with negligible emissions from mining the uranium and for manufacture of the plants. Where coal-fired power is a major component and these systems have a strong incentive to run at maximum levels, the power charging battery electric vehicles (BEVs) will be coal-fired based with high greenhouse gas emissions levels.

Where renewable energy from wind, solar, and other sources is available, this will feed directly into the grid during daytime, reducing the production from greenhouse gas-emitting power stations; and at night, when the demand is lower, this renewable electricity will contribute to recharging the vehicle batteries. In this case, the electric vehicles will be operating with a very small impact life-cycle emissions level.

The efficiency of storing electric energy in the vehicle batteries and then converting the battery energy to vehicular motion varies considerably with the type of battery. An efficiency of

70% is fairly standard; however, efficiencies of 80% or better are indicated as possible for electric energy out versus electric energy in. An efficiency of 90% is achievable, converting the electric energy to the mechanical drive by the electric motor. Regenerative braking systems improve mileage and range. Energy efficiency is considerably less, operating in cold climates by as much as 40% due to vehicle comfort heating. This can be offset to some degree, with emphasis on use of insulation. Batteries of different composition will have varying efficiencies, energy storage per kilogram, life cycle, and performance in cold climates.

The U.S. Department of Energy Pacific National Labs has indicated over 80% of existing vehicles could be converted to electric without new grid infrastructure.

The U.S. EIA has reported there were 50,000 FEVs in the United States in 2004. In 2009, the U.S. government established a target of one million by 2015. Similar targets have been established in the European Union with Spain setting a target of one million by 2014. Spain will rely on its concentration of wind energy to supply much of the electric power for its FEVs.

ELECTROSTATIC CAPACITORS

Electric double-layer capacitors are in development, which show potential to provide much higher energy density, unlimited life spans, and none of the environmental issues of regular batteries.

The University of Maryland Nanocenter is developing electrostatic nano-capacitors which increase the energy storage density by a factor of ten over that of commercially available devices, becoming competitive with electrochemical capacitors for electrical energy storage.

SOLID STATE STORAGE

Chemical storage of hydrogen in solids can provide high volume storage densities. "The most promising material systems being currently available in technical scale are complex hydrides such

as doped sodium alanates." Ref. Warner Antje, Inga Utz and Marc Linder, German Aerospace Center. The type of solid may vary depending on the nature of storage required, e.g. relative to the rate of release from storage. Cerium-doped NaAlH4 is one such material.

SUPERCONDUCTING MAGNETIC ENERGY STORAGE

Energy is stored in the magnetic field from a direct electric current in a superconducting coil cooled to a temperature below its superconducting critical temperature. The current will not decay after the coil is charged; storage of the energy is indefinite. Electricity in to electricity out has an efficiency in the order of 95%. Material costs currently limit commercial use.

WORLD HYDRO ENERGY STORAGE

World Hydroelectric Power

The total installed hydro power in the world in 2005 was 715,000 MW, representing 19% of the total electrical capacity and 63% of the world's renewables which are GHG emissions free.

Hydro dams and generating systems are often in rugged terrain distant from population load centers, requiring long high-voltage transmission lines. Weather can sometimes adversely affect the operation of the transmission lines.

Utilization of Hydro Storage

Thermal power generating plants are more often located close to load centers with the fuel being carried from the production source by rail in the case of coal and by pipeline in the case of natural gas.

During low nighttime load demand, when the thermal plants are at much reduced operating levels, the real cost of this power is only the variable cost which is the fuel element. Where daytime costs may be in the range of six cents up to 12 cents per kilowatt-hour, the real nighttime cost, when there is no demand for much of the plant capacity, will be only one or two cents per kilowatt-hour. The utility operator now has the option, if it has the luxury of a significant hydro facility, to buy this low-cost thermal power and hold back the water behind its dams, i.e., storing its own electrical energy. In the daytime, it then uses its nighttime water-stored energy and may even sell some of it to the utility with the thermal plants at the high daytime prices.

The same financial principles may similarly apply to the advantage of a utility with reasonable cost alternative energy storage.

Observations and Conclusions

- There are several technologically proven and financially acceptable methods of storing electricity during low-demand periods for delivery to the grid during peak periods.
- Sound analysis by electrical utilities of all cost factors relative to their high—and low-load periods is necessary to provide a real financial measure of the value of energy storage. This includes the avoidance of investment in GTD facilities required only to supply peak loads.
- Development of energy storage systems to remove the variability of renewable energies and provide higher value power on demand is essential for the new energy era.
- Energy storage has been essential over the ages commencing with the invention of fire and accumulating wood. Over the last century, its financial viability has justified hundreds of trillions of dollars (today's dollars) of investment. Significant investments are now necessary to achieve the full value of renewable energies.

Chapter 15

WATER USE, WATER SAVED WITH ELECTRICITY SUPPLY

A Partial Answer to the World's Water Crisis

Water Shortage

"Global warming is drying up rivers that provide water to residents of China, India, West Africa and the Southwest U.S., threatening food supplies, scientists said." Ref. Bloomberg, Center for Atmospheric Research, Van. Sun. April 2009.

Stream flows in 45 of the world's largest rivers, such as the Yellow River in China and India's Ganges, fell in the last 50 years. Ref. Study by the National Centre for Atmospheric Research (NCAR) in Boulder, Colorado.

Annual freshwater discharge into the Pacific Ocean fell by about 6% and into the Indian Ocean by about 3% (1948-2004). Annual river discharge into the Arctic Ocean rose about 10%. The flow of fresh water affects ocean circulation which helps regulate Earth's climate. Ref. Aiguo Dai, NCAR report.

Global warming from fossil-fuel emissions will reduce rain and snow runoff 10 to 30% by 2050, and water shortfalls in the Southwestern U.S. may occur in one of every three years beginning in 2010. Ref. Tim Barnett, Scripps Institution of Oceanography, San Diego.

The Financial Times of London wrote, "Water, like energy in the late 1970's, will probably become the most critical natural resource issue facing most parts of the world by the start of the next century" (year 2000).

The World Bank reports worldwide demand for water is doubling every 21 years, more in some regions.

Since 1900 there has been a six-fold increase in water use for a two-fold increase in population. This relates to rising standards of living.

Nuclear reactors, which are a large volume user of once-through cooling water, must decrease power production if the water source exceeds certain temperature or water-level requirements depending on the individual plant's operating permit. As an example, the Tennessee Valley Authority Browns Ferry Nuclear Plant in Athens, Alabama, cut back power in the summer of 2008 to keep river temperatures from exceeding environmental limits. Ref. *www.nei.org*.

A GAO 2003 report indicates 46 of the United States expect water shortages during the next 10 years under drought conditions.

WATER USAGE

The high energy and water consumption in the United States (4% of the world's population and 25% of the world's energy consumption) defines a solution or partial solution to water shortage for the United States to be as effective or more effective in China and other parts of the world which are experiencing water shortage.

The second highest use of water in the United States is associated with thermal and nuclear power generation. Planned expansion of nuclear power generation in the United States will increase its demand for water significantly. Water requirements of nuclear power plants is greater per megawatt-hour than for fossil-fueled thermal power stations due to its lower steam temperatures. Power generated from natural gas in combined cycle gas turbine (CCGT) plants has a much lower ratio of water use per unit of power generated than conventional thermal power plants.

In the United States, the highest use of water is for agriculture. The present water supply for agriculture is being used more effectively with increased application of methods, such as drip irrigation, and by reducing long-range spraying and unlined earth canals.

The U.S. Geological Survey, year 2000, reported water use in the United States at 408 billion gallons per day (Bg/d) with thermal power using 195 Bg/d or 48% of all fresh and saline water. (See note below). It noted that this level of consumption by the thermal power industry was relatively stable since 1985. Other users were irrigation at 34%, public supply at 9%, industrial at 6%, the remaining 3% being mining, livestock, and commercial.

Note: While the geological survey reports this high 48% use of water by the thermal power industry, it should be noted that 97% of this is once-through cooling water. The water is drawn from a river, large lake, or the ocean; passed through the plant condenser to remove heat from the steam turbine exhaust; and this water is then returned at a higher temperature to the same body of water from which it was drawn. Each thermal power plant has environmental limits on temperature and water levels. Other thermal power plants utilize cooling towers which remove the heat by evaporation.

WATER USE IN THERMAL AND NUCLEAR POWER PLANTS

Water is used most commonly in thermal, nuclear, and combined-cycle gas turbine (CCGT) plants as a cooling medium, condensing and cooling the exhaust steam (in a condenser) from the turbine driving the generator. Several methods are used to condense this exhaust steam. One is through the use of cooling towers where the cooling water through the steam condenser is recirculated. The most common system is with once-through cooling water, drawn from a natural source and returned to it.

The water supply to cooling towers must be of relatively good quality to avoid fouling of the condenser and of the cooling tower. The water supply is usually freshwater. Some limited use is being made of treated municipal sewage for cooling towers after osmosis-membrane purification to remove phosphates and nitrates which will foul the cooling tower due to algae growth.

The most common method of condensing the exhaust steam is by circulating cooling water through the condenser where its temperature is raised by the heat transfer. The heated water may be pumped to a cooling tower where it is sprayed down through a latticed structure, with fans drawing ambient air up through the structure. This is referred to as a closed system. The heat is dissipated by evaporation of a large portion of the water to the atmosphere. Due to evaporation, the cooling water circuit must continually be replenished with new water. The cooling tower systems use a relatively large quantity of water.

An alternative method of removing the heat from the water circulating through the steam exhaust condenser is with once-through water. This is drawn from a large lake, river, or from the ocean. The volume of water required is much larger than for the closed system, especially where the water source is already at a fairly high temperature. There is some increased evaporation from the body of water into which the once-through water is discharged, but this is relatively minor compared to the cooling tower system. The temperature rise does require monitoring to avoid damage to aquatic life. Depending on the quality of the water being drawn, there can be significant cost involved in treating it sufficiently so as not to excessively foul the steam condenser.

The table following indicates the relative use and consumption of water with the two main exhaust steam condensing systems.

Cooling Water Withdrawal and Consumption
For Thermal Power Plants (P.R. Freedman, J.R. Wolfe, Oct. 2007)

Cooling System Type	Water Withdrawal (gal/MWh)	Water Consumption (gal/MWh)
Once Through Cooling	20,000 to 50,000	Approx. 300
Cooling Towers	500 to 600	Approx. 480

Another alternative for removal of the heat from the steam turbine exhaust in the thermal power plant is the air condenser. The exhaust steam passes through a bank of finned tubes, and fans circulate air across these tubes, condensing the steam. The cost of these systems adds 5-10% to the total plant investment cost, depending on the region's ambient temperature in which the plant is located. There is also a significant increase in the parasitic power load of the plant due to the energy requirement of the fans being considerably more than that of the cooling water pumps and fans in the closed water system. This reduces the net electric power output of the plant. Both factors increase the cost of the electricity.

There are other water uses in thermal power plants, which are not significant compared to the evaporation loss in cooling towers. These are boiler blowdown (required to avoid build-up of chemicals), cooling tower blowdown (to avoid chemical and minerals build-up), and drift loss from the cooling tower.

A report from the Australian Parliamentary Library, December 2006, states that fossil-fueled steam power plants with cooling towers have a water consumption of 480 U.S. gallons per MWh compared to nuclear steam power plants with cooling towers of 720 U.S. gallons per MWh.

Water Usage By Renewables With Partial Hydrogen Storage

When evaluating an electrical power supply system from renewables—such as wind and solar—in comparison with thermal and nuclear power facilities, the relative volume of water required is becoming an increasingly significant factor, especially as climate change with increasing prevalence of droughts is reducing the natural sources of water.

The renewables of wind and solar can supply the electricity grids to meet demand; and when the supply exceeds the demand, the excess can be converted to hydrogen, stored, and returned as electricity to the grid as needed through fuel cells. Water is required in many forms of electrolysers used to convert the electricity to hydrogen. Some water is recovered when the hydrogen is combined with oxygen from the atmosphere in the fuel cell operation. There is a net consumption. This will be compared with the use of electricity from thermal and nuclear sources.

The private Idaho hydrogen plant (chapter 6) reports water usage of 2.48 U.S. gallons per hour with a 91% plant capacity factor (7,972 hours/year) for a total of 19,771 U.S. gallons/year.

Gross electricity input is 650,000 kWh/yr. or 650 MWh/yr. This indicates a gross water usage of 30 U.S. gallons/MWh. The plant is producing 8,614 kg of hydrogen annually. At a molecular weight ratio of hydrogen to water of 1 to 9, this would indicate a direct electrolysis water requirement of

>9kg H_2O/kg H_2 x 8,614 kg/yr H_2 = 79, 626 kg.H_2O/yr
>One U.S. gallon of 16 lbs./2.2 kg/lb. = 3.636 kg./U.S. Gal.
>79,626 kg H_2O/yr /3.636 kg/U.S. gal = 21,899 U.S. gal/yr.
>21,899 U.S. gal./yr/ 650 MWh = 33.7 U.S. gal/MWh.

Gross water usage may be in the order of 35 U.S. gallons per MWh of electricity into the electrolyser.

Norsk Hydro report feed water consumption in an electrolyser of one liter per Nm^3 of hydrogen.

>1 L H_2O/Nm^3 H_2 x 11 Nm^3 H_2/kg H_2 / 3.785 L / U.S. gal
>H_2O = 2.9 U.S. gal H_2O / kg H_2

Norsk report electricity consumption is 4.1 to 4.4 kWh per Nm^3 of hydrogen depending on their electrolyser model and degree of compression;

>4.4 kWh/Nm^3 H_2 x 11 Nm^3 H_2 / kgH_2 = 44 kWh/kgH_2
>2.9 U.S. gal H_2O / kg H_2 / 44 kWh / kgH_2 = .066 U.S. gal /kWh
>(or 66 U.S. gal. H_2O / MWh.)

Global conversion efficiency from electric power into the electrolysers to electric power out of the fuel cells is currently in the order of 25% with an achievement target of 40%. A measure of the water required per MWh out of the fuel cells will therefore be 4 times the electrolyser water requirement or 4 x 35= 140 (Norsk 4 X 60) gallons per MWh. The net will be less by about 20 to 25% when water from the fuel cell operations is included. At a 40% facility efficiency, the gross water requirement would be 2.5 x 35 = 87.5 (say 90) gallons per MWh. On the assumption that 40% of the renewable energy is supplied directly to the grid, there will be no water requirement for that portion. The *average* water consumption would be 60% of 90 = 54 gallons per MWh.

Water Saving by Renewables Replacing Thermal and Nuclear Generation

In the United States, one-half of its electrical production is from conventional thermal power plants; most of these are coal fired.

Electrical Production in the United States for 2006

Ref. Wikipedia

Power Source	Units in Operation	Nameplate Capacity MW	Annual Production (billion kWh)	Percent of annual Production
Coal Fired Boilers	1,480	333,115	1,995	49.1
Nuclear Power	104	105,584	787	19.4
CCNG	1,686	216,269	505	12.4
Hydroelectric	4,138	96,988	282	7.0
Renewables				
Biomass	270	6,256	53.5	1.3
Wind Power	341	11,603	30.3	0.7
Geothermal	215	3,170	13.5	0.3
Solar Energy	31	411	2.1	0.1
Sub-total (ex. Biomass & Geothermal)			32.4	0.8
Other	8,639	301,778	262.5	9.6
Total			3,930.9 billion kwh (approx. 4 billion MWh.)	

Water Consumption by Existing U.S. Power Generation (2006)

Power Source	Annual Production (billion kWh)	Water Consumption Gallons per MWh	Total Annual Water Consumption Gallons
Coal Fired Boilers	1,995	0.97% Once Through 300 gal/MWh 0.03% Cooling Towers @ 480 gal/MWh	1.995 billion MWh x 0.97 x 300 gal/MWh=580 billion gallons/yr. 1.995 billion MWh x 0.03 x 480 gal/MWh=29 billion gallons/yr.
Nuclear Power	787	720 gal/MWh	0.787 billion MWh x 720 gal/MWh=567 billion gallons/yr.
CCNG (CCGT)	505	120 gal/MWh	0.505 billion MWh x 120 gal/MWh=61 billion gallons/yr.
Biomass & Geothermal	67	Assume using cooling towers 480 gal/MWh	0.067 billion MWh x 480 gal/MWh=32 billion gallons/yr.
Renewables-non-thermal	32.4	Electrolyser water @ 54 gallons/MWh	.0324 billion MWh x 54 gallons/MWh=2 billion gallons/yr.
Others	262.5	Cooling system not avail. May be air-cooled, e.g., Diesel.	Not included
Totals	3,931		1,271 billion gallons/yr See Note following.

Note:
This would be 3.482 billion gallons per day for power generation in the United States. The U.S. Geological Survey for year 2000 reported 195 Bg/d. If once-through water is treated as "use" with approximately 35,000 gal/MWh for 0.97% of 1.995 billion MWh/yr, this would indicate 67,730 billion gallons/yr or 186 Bg/d. The 3.482 Bg/d represents consumption, whereas the U.S. Geological Survey includes all once-through water as "use."

Many of the United States have mandated a target for renewables to be 20% of total electricity requirement by 2020. The following assessment is based on 20% of all electrical production in 2020 in the United States being supplied by renewables. It assumes an increase in electric power usage by 2020 at an annual rate of 1.5% (29% cumulative from 2006 to 2020. Biomass is not included in the table following as the proportion used in thermal plants versus gas production and other methods of energy recovery has not been assessed. CCNG plants have not been assessed.

Water Consumption in 2020 with Renewables

At 20% of Total Electricity Production and the Balance of New Generation from Coal-Fired Plants for Simplicity of Comparison (see note following)

Coal-Fired Plants	Total Prod. In 2006 MWh Billions	Projected Prod. 2020 Billion MWh	Water Use per MWh (returned)	Water Consumption per MWh	Total Annual Water Consumption 2020. Billion Gallons
a. Once-through cooling (97%)	1.935	2.363	20,000 gallons (favorable)	300 gallons	709
b. Cooling Towers	0.060	0.073	n.a.	480 gallons	35
Nuclear Power	0.787	0.787		720 gallons (assumes cooling towers)	567
CCNG (CCGT)	.505	0.505		120 gallons	61
Wind/Solar (60% stored by electrolysis)	.0324	Projected Prod. @ 20% 1.014 billion	n.a.	54 gallons	55
Biomass & Geothermal	.067	0.067		480 gallons	32
Others	.2625	0.2625		May be air-cooled, e.g., Diesel	n.a.
Total Production in the United States	3.931 billion	5.071 billion			1,459

Note to table: If nuclear power production by 2020 provides some of the new generation requirements, reducing the proportion from coal-fired power; this will further increase the cooling water consumption by a ratio of 700:300.

With no increase in renewables, then if new coal-fired plants were built to supply *all* this electricity increased generation of 1.014 billion MWh/yr by 2020, and assuming all cooling is once through, there would be added water consumption of 1.014 billion MWh/yr x 300 gal/MWh = 335 billion gallons/yr. The equivalent renewables would have used 55 billion gallons/yr.

In the case of increased coal-fired power generation, the existing standards of water use have been applied. However, if concentrated carbon sequestration (CCS) is applied for new coal-fired power generation, the water consumption would increase by 10-20% due to the additional energy consumed to achieve concentrated carbon dioxide and for subterranean sequestration.

A comparison has not been provided here for water requirements of oil refineries producing gasoline and diesel from crude oil with the water requirement of electrolysis production of hydrogen for fuel cell vehicles. Considerable water savings would occur with replacement of gasoline and diesel by hydrogen for vehicles.

OBSERVATIONS AND CONCLUSIONS:

A benefit of increased generation of wind and solar renewables is the large reduction in consumption of water, which is becoming an increasingly critical commodity.

New electricity supply from renewables of wind and solar will be more practical and acceptable than thermal or nuclear plants in regions with water shortages or temperature limits for once-through cooling.

In some regions permitting for new thermal and nuclear plants may not be possible due to the large water requirements.

Chapter 16

OBSERVATIONS AND CONCLUSIONS

AN OVERWHELMING AND IRREVERSIBLE CHANGE IN PROGRESS

WAVES OF INNOVATION

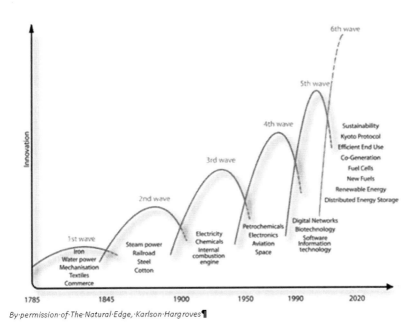

By permission of The Natural Edge, Karlson Hargroves

It requires more knowledge, better visions of the future and an unlimited ability to change.

Essential change does not come easily when powerful special interests may be affected.

1876

A Boston newspaper editorial wrote about the telephone: "Well-informed people know it is impossible to transmit the voice over wires, and that if it would be possible to do so, the thing would have no practical value."

1878

A British professor—after viewing the electric light at a world science exposition—remarked, "When the Paris exhibition closes, the electric light will close with it, and no more will be heard of it."

1898

Charles Duell, the commissioner of the U.S. Office of Patents, urged President McKinley to abolish his office, arguing, "Everything that can be invented has been invented."

Observations

Renewable energies will be the clean fuels of the future with energy storage as hydrogen along with other efficient proven means.

Current hydrogen production from natural gas produces excessive GHGs.

Today's emphasis on vehicle emission reductions with hydrogen highways provides the infrastructure to use clean hydrogen in the future.

Clean hydrogen from renewable electricity at competitive costs can be achieved with university and industry research and development of electrolysis, fuel cells, and infrastructure.

National hydrogen energy research institutes in all the regions which are moving forward with renewable energies can lead in developing a clean hydrogen economy.

Conclusions

Costs Will Be Reduced by Mass Production to a Lower Level than the Competition

Renewable energies have shown dramatic reduction in costs with major increases in efficiency. Consistent longer-term policies and incentives by government is essential for industry to be able to develop and expand renewable energy generation facilities, moving into the mass-production level which will provide major cost reductions. Short-term policies and incentives are a deterrent to reaching the mass-production low-cost era for renewables.

Progress with Renewables

Significant annual growth of installations for wind energy and solar power has been reported for some years from several countries. In Europe, there have been enlightened policies, especially with feed-in tariffs, which have resulted in renewables becoming a significant portion of their total energy mix. China has made a major leap forward to achieve, by far, the greatest annual additions of wind and solar power of all countries. They have become the leaders in both installations and manufacturing. China's move to consolidate the manufacturing sectors in these fields will have a major favorable impact on their unit costs. It is clear that China listened carefully to the speech in Beijing by Thomas Friedman, described in his book *Hot, Flat, and Crowded*.

Energy Storage

There are a number of proven efficient methods of storing energy. Developing these facilities is a necessary component of acceptable major growth of variable renewables.

Investment in Renewable Energies

Accelerating the development of renewable energies does require appropriate government policies and incentives. These need only represent a fraction of the investment cost as developers will be able to provide the major component of capital. Such incentives will only be needed until the rate of development achieves the mass production level.

Electricity Costs

Most electrical utilities have their return on investment and rates set by Public Utility Commissions. There is no real incentive to use vision in planning for the greater benefit of the public.

Investment in research and development by electrical utilities shows only a fraction of that committed by major industrial businesses. Some electrical utilities have show enlightenment by establishing non-regulated divisions.

It is necessary for electrical utilities to develop the real cost of peak power and of low-demand period power. These will serve to justify energy storage and distributed power projects.

Eliminating Overseas Oil

In the countries of the developed world rapid growth of the infinite wind and solar energies to eliminate the finite overseas oil can be achieved with only a fraction of the cost of that oil applied as incentives. The major portion of the investment will be supplied by the developers. Eliminating the vast cash flow to purchase finite overseas oil, which is sure to increase in cost with world-wide increasing demand, will improve the national economy, reduce political tensions and military demands.

Reducing Greenhouse Gas Emissions

Major development of renewable energies will displace the output of GHG emitting fossil fuel generating plants; thereby contributing in a major way to reducing the threat of climate warming.

Critical Water Shortage

Some regions of the world are encountering a continuing reduction of available water. Renewable energies replacing fossil fuel and nuclear generating plants provide a major reduction in the need for water.

Urgent National Needs (United States)

"I believe we possess all the resources and talents necessary. Bur the facts of the matter are that we have never made the national decisions or marshalled the national resources required for such leadership. We have never specified long-range goals on an urgent time schedule, or managed our resources and our time so as to insure their fulfilment" U.S. President John F. Kennedy, a message to congress on "Urgent Nation Needs, May 25, 1961.

INDEX

A

airplane, 3
alternative sources of energy, 5
ammonia, 8, 77, 120, 138-41, 157, 172-73
anhydrous ammonia energy storage, 172

B

Becquerel, Edmond, 50
Benzene, 8
Berman, Elliot, 50
bioenergy, 7, 69
biomass, 34, 108, 124-25

C

canada, 77, 90, 109, 111
   hydroelectricity in, 5
   wind energy developments in, 34
capacity factor (CF), 110, 124
carbon capture and sequestration (CCS), 8
carbon dioxide, 6, 15-16, 39, 52, 141, 145, 169
   credit purchase of, 95
   emissions, 90, 92, 99
   exhaust and sequestration of, 8
   molecular weight of, 97-98
carbon-neutral supply of energy, 7
carbon-to-nitrogen ratio, 100
C.D. Howe Institute, 44
Center for Energy and Economic Development, 87
chemical catalyst, 11
Chemical Engineers Handbook (Perry), 10
china, 193
   electricity consumption of, 126
   greenhouse gas emissions of, 90, 93
   vehicles in, 143, 155
   water shortage in, 179
   wind energy installations in, 25-26, 30, 47
climate change, 7, 91, 99, 173, 183
climate warming, 100, 133
coal, 2-3, 5-6, 71, 96, 135, 155, 168, 174, 176
   cost of, 46, 112
   and hydrogen production, 18
   power generation from, 111
   power plants, 118, 147, 185
combined cycle gas turbine (CCGT) generating plants, 114
combined heat and power (CHP), 45

compressed air energy storage (CAES), 163-64, 166-67, 169
Compressed Air Vehicles (CAVs), 148
concentrator photovoltaic (CPV), 59
crude oil, 131, 152, 157
  as most versatile fossil fuel, 4
  replacement by hydrogen, 12

D

"Deploying Renewables: Principles for Effective Policies," 12
deserts, potential of, 46
distributed power centers (DPCs), 13, 103-8, 110-13

E

Einstein, Albert, 50
electricity, 2, 6, 10-11, 13, 27, 29-30, 32
  Chinese usage of, 126
  consumption of, 5, 22
  demand curve of, 105
  generation of, 23
  storage of, 159
  U.S. usage of, 126-27
electric system operators (ESOs), 160
electric vehicle energy storage, 173-74
electric vehicles (EVs), 2, 46, 154-55, 161, 173-74
  electricity consumption of, 152
  motives for conversion to, 153
electrolyser, 11, 70-72, 75

electrolysis, 7, 18, 45, 68, 109, 132, 134, 192
  efficiency of, 11, 13, 32, 52, 70-77, 81-83, 106, 113, 120, 122-23, 132, 183-84
  and hydrogen conversion, 11, 68, 75, 80, 88, 95, 122, 124, 138
  in solar-powered hydrogen fueling station, 75
  theory, 10
electrostatic capacitors, 157, 175
emissions, car, 144
Energy Storage, xi, 69, 119, 157, 159, 177
  advantages of, 160
  flywheel, 163
  wind energy systems and, 42, 111
energy systems, sustainability of, 119
Ethyl alcohol, 8
European Environmental Agency (EEA), 92
european union, 69, 90, 92, 119, 175
European Wind Energy Association (EWEA), 31
European Wind Energy Council (EWEC), 34

F

faraday, 10
feedstocks, 124
foreign oil, 131, 133
fossil fuels, xi, xiii-xv, 2-8, 15, 69, 100-101, 119
france, 55, 174

Friedman, Thomas
Hot, Flat, and Crowded, 193
fuel cells, 71, 103-4, 106-8, 113, 183-84, 192
   electricity generation through, 11, 19, 32, 52, 74-75, 173
   emissions from, 145
   and hydrogen energy production, 16, 51, 126-27
fuel cell vehicles, 47-48, 128-29, 142, 145, 189

G

gasoline, 13, 47, 69, 95, 133, 148, 151-52, 157
   compared to hydrogen, 135
   and electric vehicles, 129, 152, 155, 173
   and greenhouse gas emissions, 135, 144
   U.S. consumption of, 128-29, 131
generation, transmission, and distribution (GTD), 103, 106, 112
geothermal energy, 34
German Advisory Council on Global Change (WBGU), 25
Global Wind Energy Council (GWEC), 135
green house gases (GHGs), 47, 99
   beef cattle contribution to, 96
   emissions, 90, 92-93, 96, 98, 100, 104, 144-45, 147
   reduction of, 135
"Green Path from Fossil-Based to Hydrogen Economy," 7

H

Haber-Bosch process, 138
high-concentration photovoltaics (HCPV), 59
Higher Heat Value, 11, 126
Hot, Flat, and Crowded (Friedman), 193
hydroelectricity, 5
hydro energy, 34, 69
   sources of, 3, 5
hydrogen, 3, 6
   buffer system, 32-33
   compression of, 16
   conversion of, 18, 103, 129, 151
   current production of, 7
   and electrolytic chemical plants, 19
   as gasoline substitute, 133
   Idaho production plant, 80-81, 83
   others sources of, 87
   process efficiency of, 15
   production and conversion of, 15, 17
   safe use of, 9
   uses of, 95, 119-20
   as vehicle fuel, 151
hydrogen and vehicles, 143, 145
hydrogen fuel cells, 120, 127
   vehicles with, 128, 142
hydro storage, 170-71
hythane, 18, 119

I

india, 66, 179
innovation, waves of, 2
internal combustion engine (ICE) vehicles, 2, 19, 74, 104, 119, 127, 129, 139, 144-45, 151, 153-55, 173

International Energy Agency
(IEA), 12, 69
International Energy Outlook
2009, 30
Italy, 64

J

Japan, 161
 photovoltaic installations in, 56, 65
 solar power developments in, 51
joules, 9, 21, 126-27, 203

K

Krynski, Jami, 40
 "On Renewable Energy Technologies," 40

M

Massachusetts Institute of Technology (MIT), 11
methane, 8, 15-18, 78, 94, 97-98, 103, 125, 144, 204
molten salt energy storage, 172
Muradove & Veziroglus, 5
Mysterious Island, The (Verne), 10

N

National Hydrogen Association (NHA), 87
National Renewable Energy Lab (NREL). *See* U.S. DOE National Renewable Energy Laboratory (NREL)
natural gas, xiv-xv, 3, 5, 15, 17-18, 23-24, 36, 94-95, 107-8, 119-20, 138-39, 141, 163-64, 168-69
Nocera, Daniel, 11
Norsk Hydro Electrolysers, 73, 120, 122-25, 138, 184
nuclear power, 5, 7, 40, 87, 185-86
nuclear sources, 3

O

off-peak generators, 71
"On Renewable Energy Technologies" (Krynski), 40
organic agriculture, 99

P

peak demand curves, 104-7
Perry, John
 Chemical Engineers Handbook, 10
photovoltaic (PV) conversion efficiency, 58
photovoltaic (PV) panel, 50
photovoltaics, 50, 124
portugal, 64
pressure swing adsorption (PSA), 16-17
PUMPED HYDRO ENERGY STORAGE (PHES), 170-71

R

renewable energy, 3-4, 6, 193
 advantages of, xi, 118
 in Canada, 40
 in Europe, 31-32

potential of, 121
sources of, 34
variables of, 69, 109

S

satellites, 3, 121
slovakia, 65
solar energy, 34, 50, 57, 60, 63, 66-67, 126, 138, 185, 194
  developments in, 60, 62, 64-65
  power generation from, 50, 52
  projected growth of, 57-58
  and the U.S., 124
Solar Industry, 59
solar photovoltaic systems, 51, 53-55, 59
solar thermal systems, 51, 59, 67
solid oxide electrolyser cell (SOEC) technology, 76
Space Travel, 3, 121
spain, 87
  electric energy storage in, 158
  molten salt energy storage in, 172
  photovoltaic installations in, 56, 60, 65
  renewable energy in, 32
  solar power in, 65
  wind energy in, 34, 175
Stanford University, 25, 121
steam engine, 2
superconducting magnetic energy storage, 176

T

taiwan, 66
terrawatts, 4-5, 21

U

United Nations Food and Agriculture Organization (UN FAO), 96
United States
  biomass resource in, 124
  electricity consumption of, 126-27
  solar power in, 64, 124
  wind energy in, 36-40
  wind resource in, 123, 128
University of Minnesota, 76
U.S. DOE Energy Information Administration Outlook 2009, 21
U.S. DOE National Renewable Energy Laboratory (NREL), 9, 11, 13, 37, 44, 70-76, 120-25, 132

V

vatican, 65
vehicles
  electric, 147, 152, 154
  fuel cell, 47, 145, 189
  fuel consumption of, 128-29
Verne, Jules, 10
Mysterious Island, The, 10

W

water
  shortage of, 179-80, 189
  usage of, 179, 180-89
wind energy, 12-13, 110, 113-14, 134-35, 140, 154, 158, 162, 170, 175, 193

in Canada, 34, 41
in China, 30-31, 47
costs of, 40
in Europe, 31-32, 34, 47
global installations for, 25, 27, 29
output vs. demand, 42
potential, 24, 121
in the U.S., 36-37, 39, 123, 128, 132-33

wind farms, 75, 77, 122, 161
  capacity factor of thermal/nuclear facilities, 110
  capital cost of, 41
  in Europe, 31-32
  land requirements for, 47
  requirements for, 134
  in the U.S., 36

wind-hydrogen-diesel energy system, 74
wind-hydrogen pilot system, 71
wind-to-hydrogen (W2H) development, 44, 45-46, 74
wind turbine generators (WTGs), 110, 134
world electricity generation, 22, 171
world hydro energy storage, 176

APPENDIX

UNITS AND CONVERSIONS

Units of electrical energy:

1 watt hour	= 3.413 btu/hour
1 kilowatt hour	= 3.6 x 106 joules = 3,413 btu/hr
1 mm btu	= 293 kilowatts (approx 300 kW)
1 mm btu	= 1.055 gigajoules
1,000 watts (thousand) (103)	= 1 kilowatt (KW)
1,000,000 watts (million) (106)	= 1 megawatt (MW)
1,000,000,000 watts (billion) (109)	= 1 gigawatt (GW)
1,000,000,000,000 watts (trillion) (1012)	= 1 terawatt (TW)
1,000,000,000,000,000 watts (quadrillion) (1015)	= 1 petawatt (PW)
1018 watts	= 1 exawatt (EW)

Mj – mega-joules, joules are a measure of energy, mega is million.
Btu – British Thermal Units – a measure of energy

HYDROGEN AND METHANE WEIGHTS AND VOLUMES

HYDROGEN

Weight/volume
0.0052 lbs/ft3.
1 lb. = 192.3 ft3.
1 kg = 192.3 ft3/lb x 2.205 lbs/kg = 424 ft3.
1 kg. = 424 ft3/kg / 35.31 ft3/m3 = 12.001 m3.
1 m3 = 1 m3 / 12.001 m3/kg = 0.0833 kg.

Properties   Heat of combustion 33,887.6 cal/gm (60,937.7 btu/lb) (134,062.94 btu/kg)
1 kg. of hydrogen equivalent to 1 U.S. gallon of gasoline.
1 gallon of gasoline, depending on quality, 130,000 to 135,000 btu

METHANE

Weight/volume   .04163 lbs/ft3
1 lb. = 24 ft3
1 kg = 24 ft3/lb x 2.205 lbs/kg = 52.92 ft3/kg.
1 kg = 52.92 ft3/kg / 35.31 ft3/m3 = 1.499 m3/kg
1 m3 = 1 m3 / 1.499 m3/kg = 0.667 kg.

Edwards Brothers, Inc.
Thorofare, NJ USA
May 13, 2011